Studies in Neuroscience, Psychology and Behavioral Economics

Series editors

Martin Reuter, Bonn, Germany
Christian Montag, Ulm, Germany

More information about this series at http://www.springer.com/series/11218

Christian Montag · Martin Reuter
Editors

Internet Addiction

Neuroscientific Approaches and Therapeutical Interventions

Springer

Editors
Christian Montag
Department of Psychology
Molecular Psychology
University of Ulm
Ulm
Germany

Martin Reuter
Department of Psychology
Personality Psychology and Biological Psychology, Center for Economics and Neuroscience (CENs)
University of Bonn
Bonn
Germany

ISSN 2196-6605 ISSN 2196-6613 (electronic)
Studies in Neuroscience, Psychology and Behavioral Economics
ISBN 978-3-319-34681-6 ISBN 978-3-319-07242-5 (eBook)
DOI 10.1007/978-3-319-07242-5

Springer Cham Heidelberg New York Dordrecht London

Softcover reprint of the hardcover 1st edition 2015

Printed on acid-free paper

Springer International Publishing AG Switzerland is part of Springer Science+Business Media (www.springer.com)

Preface

Dear practitioners, scientists, and students,

We are very happy that the book "Internet addiction: Neuroscientific approaches and therapeutical interventions" is finally available to you. Current developments in the DSM-V underline the importance and the need for a scientific book on this newly emerging disorder, as the term "Internet Gaming Disorder" now appears among the emerging measures and models in Section III. It has been included in this section with the aim of better understanding the nature of Internet addiction and potentially including it as a distinct disorder in the next update of DSM.

To precisely describe and characterize Internet addiction (and the related term "Internet Gaming Disorder"), strong research efforts have been invested around the globe. In this book, we give you an extensive overview of Internet addiction by leading researchers in the field. It is fantastic that such a large number of scientists doing the most important work on Internet addiction have agreed to write reviews on their research approach. We are very grateful for this support.

We are sure that the present book will not only be an excellent guide to Internet addiction for scientists and therapists working in this field, but to all individuals interested in this new topic, or even patients affected by this emerging disorder. Clearly, the topic of Internet addiction is of special relevance around the globe in this increasingly connected online world.

Of note, the book is written for students, practitioners, and researchers alike, as a wide range of different approaches to the understanding of Internet addiction—ranging from molecular genetics to brain imaging—are represented in this book. Each chapter gives a short introduction to these methods, as applied to the better understanding of Internet addiction. If you require greater detail on the neuroscientific techniques such as structural MRI or molecular genetics, we refer you to the core text, "Neuroeconomics" of the series "Studies in Neuroscience, Psychology and Behavioral Economics" in which this book is also published. This core text contains an extended appendix with individual chapters on a large number of neuroscientific methods. In addition to a very good overview on Internet addiction and neuroscientific approaches in the second section of this book, a large part of this book (the third section) gives interesting and valuable insights into the treatment of Internet addiction. Experiences from both psychopharmacology and psychological counseling are presented. The latter is also considered from a cross-cultural perspective (Germany and South Korea). Finally, insights from a governmental stance aimed at tackling this newly emerging disorder are also given. We hope that this more practical section will be of special help to counselors all over the world in treating this potential new disorder.

We thank all authors for their contribution. Without their valuable time spent on this project, this book would not have been possible. Every single contributor is a well-known researcher/practitioner in the field of Internet addiction. We are very happy that they have all agreed to participate and to contribute without hesitation. Last but not least, we thank Éilish Duke for her fantastic support in improving the language of the present book.

A last note: We are aware of the controversial discussion in scientific circles on how to name problematic Internet use. Neither it is clear if problematic Internet use actually reflects an addiction nor is Internet addiction an official diagnosis in DSM-V and ICD 10. However, taking into account early studies on this topic but also the fact that the term Internet addiction still is often used and appears as quite "handy" in everyday communication of this topic, we decided to use it for this book.

Let us know if something of relevance is missing from this book. This will help us to update the book in an eventual second edition.

Best wishes

Christian Montag
Martin Reuter

Contents

Part III Therapeutical Interventions & Governmental Policies

How to read this book

The book contains three parts, with a brief first part including a concise overview of Internet addiction by Kimberly Young, as an excellent starting point. The second part gives detailed insights into different neuroscientific approaches to Internet addiction research. If you only require a brief overview over the neuroscientific literature as a starting point, the summary Chap. 8 will be of help. Chapters 2–7 deal with distinct perspectives such as structural brain imaging and Internet addiction. If you are mainly interested in therapeutic interventions in the context of Internet addiction, you can skip the second part of the book (perhaps with the exception of Chap. 5 dealing with functional changes of the brain related to the treatment of Internet Gaming Disorder) and immediately go to the third part, which provides different perspectives on the treatment of Internet addiction. Finally, the book closes with an Appendix called Neuroanatomy comprising several figures on human brain anatomy, intended to accompany the chapters of the second part. In addition, a short glossary explains key vocabulary from the book.

Part I
Introduction to Internet Addiction

Chapter 1
The Evolution of Internet Addiction Disorder

Kimberly Young

Abstract This chapter presents the history and evolution of Internet addiction and describes the risk factors identified. As the problem has become more widespread, new studies examine the neuroscientific causes of Internet addiction and ways that the disorder may be treated primarily using behavior therapy, cognitive-behavioral techniques, and residential care. The chapter also provides the theoretical frameworks to understand the etiologic models or causal factors associated with the development of Internet addiction including a brief overview of the neuroscientific studies recently done. Finally, this chapter reviews the current treatment models used in Internet addiction recovery. As an introduction to this book, it is hoped this chapter gives a historical context of the disorder and promotes future areas of research as new studies in the field continue to emerge.

1.1 Introduction

Internet addiction was first researched in 1996 and findings were presented at the American Psychological Association. The study reviewed over 600 cases of heavy Internet users who exhibited clinical signs of addiction measured through an adapted version of the DSM-IV criteria for Pathological Gambling (Young 1998). Since then, subsequent studies over the past decade have examined various aspects of the disorder. Early studies attempted to define Internet addiction and examined behavior patterns that differentiated compulsive from normal Internet usage. More recent studies investigated the etiologic factors or causes associated with the disorder. Much of this examined the impact of computer-mediated communication on the way people will adapt to interactive features of the Internet and initial studies from the United States spread into countries such as Taiwan, Russia, China, and the United Kingdom.

K. Young (✉)
Center for Internet Addiction, St. Bonaventure University, Allegany, NY, USA
e-mail: kyoung@sbu.edu

C. Montag and M. Reuter (eds.), *Internet Addiction*, Studies in Neuroscience, Psychology and Behavioral Economics, DOI 10.1007/978-3-319-07242-5_1

As the problem has become more widespread, new studies examine the neurological causes of Internet addiction and ways that the disorder may be treated primarily using behavior therapy, cognitive-behavioral techniques, and residential care. This chapter presents the history and evolution of Internet addiction and describes the risk factors identified. The chapter also provides the theoretical frameworks to understand the etiologic models or causal factors associated with the development of Internet addiction including a brief overview of the neurological studies recently done. Finally, this chapter reviews the current treatment models used in Internet addiction recovery. As an introduction to this book, it is hoped this chapter gives a historical context of the disorder and promotes future areas of research as new studies in the field continue to emerge.

1.1.1 Diagnosis of Internet Addiction

Diagnosis of Internet addiction is often complex. Unlike chemical dependency and substance abuse, the Internet offers several direct benefits as a technological advancement in our society and not a device to be criticized as addictive. Individuals can conduct research, perform business transactions, access libraries, communicate, and make vacation plans. Books have been written outlining the psychological as well as functional benefits of the Internet in our lives. By comparison, alcohol or drugs are not an integral or necessary part of our personal and professional lives nor do these substances offer any health benefit. With so many practical uses of the Internet signs of addiction can easily be masked or justified. Further, clinical assessments often cover relevant disorders for psychiatric conditions and addictive disorders. However, given its newness, symptoms of Internet addiction may not be revealed in an initial clinical interview. While self-referrals for Internet addiction are becoming more common, often the client does not present with complaints of computer addiction. People may initially present with signs of depression, bi-polar disorder, anxiety, or obsessive-compulsive tendencies, only for the treating professional to later discover signs of Internet abuse upon further examination (Shapiro et al. 2000). Therefore, diagnosing Internet addiction upon clinical interview can be challenging. It is important to understand the current definitions of Internet addiction to help treating professionals screen for patients for compulsive use of the Internet and the evolution of Internet addiction as a disorder as part of the assessment process.

1.1.2 The Evolution of Internet Addiction

While time is not a direct function in diagnosing Internet addiction, early studies suggested that those classified as dependent online users were generally excessive about their online usage, spending anywhere from 40 to 80 h/week on recreational or private use of the Internet with sessions that could last up to 20 h (Young 1998).

Sleep patterns were disrupted due to long Internet sessions where addicts often took caffeine pills to facilitate longer Internet sessions and suffered from fatigue, poor diet, poor exercise, work and/or school performance due to loss of sleep.

The best method to clinically detect compulsive use of the Internet is to compare it against criteria for other established addictions. Researchers have likened Internet addiction to syndromes similar to impulse-control disorders on the Axis I Scale in the DSM (APA 1994) and utilized various forms of DSM-IV based criteria to define it. Of the all the references in the DSM-IV, Pathological Gambling was viewed as most akin to this phenomenon. The Internet Addiction Diagnostic Questionnaire (IADQ) was the first screening measure developed for diagnosis (Young 1998) that conceptualized the criteria for the disorder as follows:

1. Do you feel preoccupied with the Internet (think about previous online activity or anticipate next online session)?
2. Do you feel the need to use the Internet with increasing amounts of time in order to achieve satisfaction?
3. Have you repeatedly made unsuccessful efforts to control, cut back, or stop Internet use?
4. Do you feel restless, moody, depressed, or irritable when attempting to cut down or stop Internet use?
5. Do you stay online longer than originally intended?
6. Have you jeopardized or risked the loss of significant relationship, job, educational or career opportunity because of the Internet?
7. Have you lied to family members, therapist, or others to conceal the extent of involvement with the Internet?
8. Do you use the Internet as a way of escaping from problems or of relieving a dysphoric mood (e.g., feelings of helplessness, guilt, anxiety, depression)?

Answers evaluated non-essential Internet usage such as for non-business or academically related use. Subjects were considered 'dependent' when answering endorsing five or more of the questions over a six-month period not associated with manic or hypomanic episodes. Associated features included neglect of routine duties or life responsibilities, social isolation, and being secretive about online activities or a sudden demand for privacy when online.

Throughout the late 1990s–2000s, the IADQ was widely used in studies on Internet addiction. As researchers adapted the IADQ, some made refinements to the instrument. Beard and Wolf (2001) recommended that all of the first five criteria be required for diagnosis of Internet addiction, since these criteria could be met without any impairment in the person's daily functioning. They also recommended that at least one of the last three criteria (e.g., criteria 6, 7, and 8) be required in diagnosing Internet addiction. These criteria impact the pathological Internet user's ability to cope and function (e.g., depressed, anxious, escaping problems), and also impact interaction with others (e.g., significant relationship, job, being dishonest with others). Other studies that empirically tested the IADQ found that using 3 or 4 criteria where just as robust in diagnosing Internet addiction as using 5 or more and suggested that the cutoff score of 5 criteria might be overly stringent (Dowling and Quirk 2009).

The most recent acceptance of Internet addiction is the inclusion of Internet Use Gaming Disorder in Sect. 1.3 of the DSM-V. Researchers had encouraged the inclusion of pathological Internet use as a disorder (e.g., Block 2008) given the volume of studies that identified it as a problem. The main concern was that various criteria had been used to diagnose and classify Internet addiction in the literature. Section 1.3 will include conditions that require further research before their consideration as formal disorders, as well as cultural concepts of distress, the names of individuals involved in DSM-V's development, and other information. As a new condition, including it in the DSM-V will provide a standardized set of criteria for future studies.

1.1.3 Internet Addiction Test

Beyond DSM criteria, the Internet Addiction Test (IAT) is the first validated instrument to assess Internet addiction (Widyanto and McMurren 2004). Studies have found that the IAT is a reliable measure that covers the key characteristics of problematic Internet use. The test measures the extent of client's involvement with the computer and classifies the addictive behavior in terms of mild, moderate, and severe impairment. The IAT can be utilized among outpatient and inpatient settings and adapted accordingly to fit the needs of the clinical setting.

The IAT is a worldwide accepted and validated testing instrument that examines symptoms of Internet addiction such as a user's preoccupation with Internet use, ability to control online use, extent of hiding or lying about online use, and continued online use despite consequences of the behavior. The IAT has been validated in France (Khazaal et al. 2008), Germany (Pawlikowski and Brand 2011), Norway (Johansson and Götestam 2004), Finland (Kaltiala-Heino et al. 2004; Korkeila et al. 2010), Italy (Ferraro et al. 2007), Greece (Siomos et al. 2008), Iran (Ghassemzadeh et al. 2008), Pakistan (Suhail and Bargees 2006), China (Lam et al. 2009), and Korea (Hur 2006). Tao et al. (2010) also proposed that a diagnostic score of 2 + 1, where the first two symptoms (preoccupation and withdrawal symptoms) and at least one of the five other symptoms (tolerance, lack of control, continued excessive use despite knowledge of negative effects/affects, loss of interests excluding internet, and use of the internet to escape or relieve a dysphoric mood) was established. This makes the IAT the first globally psychometric measure of the disorder.

1.2 Risk Factors for Internet Addiction

As Internet Use Disorder has gained credibility, more studies focused on risk factors associated with the development of the disorder. The risk factors can loosely be categorized as social factors, psychological factors, and biological factors, each will be further discussed.

1.2.1 Social Factors

Excessive or problematic Internet use often stems from interpersonal difficulties such as introversion or social problems (Ebeling-Witte et al. 2007). Often, Internet addicts fail to communicate well in face-to-face situations (Leung 2007). This is partly why they use the Internet in the first place. Communicating online seems safer and easier for them. Poor communication skills can also cause poor self-esteem, feelings of isolation and create additional problems in life, such as trouble working in groups, making presentations, or going to social engagements. Virtual relationships are a way of engaging with others while having the safety of avoiding rejection or the anxiety of making physical contact with others. Shyness can be consuming and the Internet offers an immediate relief the anxiety this causes. Therapy needs to address how addicts communicate offline and to establish positive new ways of interacting. Furthermore, in the context of the IAT, a recent study found inverse correlations between the IAT and self-directedness in a group of healthy participants from the population and first-person-shooter-video-players (Montag et al. 2011). A new study from this group shows that this effect can be found cross-cultural in seven countries (Sariyska et al. 2014).

Other research has focused on limited social support systems that Internet addicts have, which is in part why they turn to virtual relationships as a substitute for the missing social connection in their lives. They turn to others on the Internet when feeling lonely or need someone to talk with. Studies have found loneliness is associated with the development of Internet addiction (e.g., Hardie and Tee 2007; Morahan-Martin 1999). Loneliness as a risk factor is consistent with findings that suggest social relationships are a key component in the development of Internet addiction. The most addictive applications are chat rooms, interactive games, instant messaging, or social media, suggesting that the condition is socially motivated.

A more discrete social risk factor for Internet addiction is the development of online affairs (Whitty 2005). An online affair is a romantic or sexual relationship initiated via online contact and maintained predominantly through electronic conversations that occurs through email, chat rooms, or online communities (Atwood and Schwartz 2002). This again shows that interpersonal problems or loneliness can play an active role in developing an addiction to online communication and relationships. Here, marital problems and discord play a role but it is unclear to what extent. Do marital problems come first before establishing new romantic relationships online or does the anonymity of online relationships accelerate intimacy online? Or, does the accessibility of meeting others online create the opportunity for affairs to begin among otherwise stable or healthy marriages. New research is trying to address these questions.

1.2.2 Psychological Factors

There are two types of Internet addicts. The *Dual Diagnosed Internet Addict* suffers from prior psychological problems such as to depression, anxiety, obsessive-compulsive disorder, or substance abuse, to name a few syndromes associated with the disorder. Other addicts, referred to as *New Internet Addicts*, have no prior history of psychiatric illness or addiction, and their addiction to the Internet is an entirely new problem. Dual Diagnosed Internet Addicts may suffer from a variety of illnesses that contribute to developing Internet addiction whereas the New Internet Addict do not have any psychiatric history but focus on particular activities or relationships online (a specific online affair, chat room, message board, game, gambling site, or adult site, to name a few examples). The disorder exists solely online (Young 2004).

Dual Diagnosis Internet Addicts suffer from depression (Ryu et al. 2004), social anxiety (Yen et al. 2007), impulsivity (Lavin et al. 1999), obsessive-compulsive disorders (Shapiro et al. 2000), and general psychiatric problems (Yen et al. 2008). Dual Diagnosed Internet Addicts suffer from alcohol or drug dependency only to find their compulsive use of the Internet a physically safe alternative to their addictive tendency (Young 2004). They believe that being addicted to the Internet is medically safer than being addicted to drugs or alcohol; at the same time, the compulsive behavior avoids the need to confront unpleasant feelings or situations underlying the addictive behavior.

Dual diagnosis in addiction is common. The Dual Diagnosed Internet Addict can be displayed in a variety of ways unique from other addictive syndromes. Research has not confirmed which is cause and effect but we have established a clear correlation between Internet addiction and psychiatric problems. For instance, we know that Internet addicts suffer from depression but it is hard to know which came first. Some suggest that because a person suffers from depression that he or she uses the Internet as a means to cope with sad feelings and low self-esteem associated with the disorder. The person goes online to forget about sad feelings as they escape into the Internet. It is also possible that as a person goes online with increased frequency, he or she may feel more depressed as they become socially isolated from others.

In another example, a person suffering from anxiety may seek out companionship in a safe virtual environment. In yet another example, a sexual compulsive discovers a new source for sexual gratification through online pornography and anonymous sex chat. The Internet allows them to continue their sexual behavior without the physical need to visit strip clubs or prostitutes and provides a new and socially acceptable way to cope. Realizing the impact of this destructive behavior, the person rationalizes it and continues to engage in the activity despite its known potential risks, including possible job loss, divorce, or arrest. The online experience turns into a relief from pain and anxiety, the reward for success, and a way to avoid addressing other painful emotions. The online world becomes a private refuge and while the Dual Diagnosed Internet Addict progressively retreats into the computer, it is unclear how psychiatric history plays a role.

New Internet Addicts meet two distinct criteria. First, they become addicted to new forms of Internet use created solely online such as chat rooms, social networks, instant messaging, role-playing games, or eBay. Someone who becomes addicted to chat rooms must use the Internet to chat. Someone who becomes addicted to eBay must use the Internet to access it. Granted, these activities have now become portable through mobile devices such iPhones, Droids, or iPads or cell phones. The key element is that they are all considered Internet-specific activities.

Secondly, New Internet Addicts are individuals with no previous significant addictive or psychiatric history. They develop an addiction to the anonymous, accessible, and interactive nature of online use. For instance, New Internet Addicts may include a 50-year-old lawyer using sex chat rooms during work hours and without his wife's knowledge, a 30-year-old business executive compulsively checking his iPhone to check his match.com girlfriends, a 20-year-old college student constantly uses Facebook, or a 16-year-old boy constantly playing World of Warcraft with no other comorbidity. The compulsive Internet use is a new clinical phenomenon.

1.2.3 Biological Factors

The most recent research reflected in this book focuses on the biological studies associated with Internet addiction. I won't go into as much detail in this section as this book provides substantial depth in this new area of research. I will highlight a few notable studies that are very helpful to learn what causes Internet addiction. We know that psychological and social factors are associated with the development. We have little knowledge about the biological associations with this disorder. This is why these are important new studies that allow us to see the biology of Internet addiction and more broadly, the biology of addiction in general.

We have learned through functional magnetic resonance image (fMRI) analyzing the differences between addicts and non-addicts that brain regions such as the cerebellum, brainstem, right cingulate gyrus, bilateral parahippocampus, right frontal lobe (rectal gyrus, inferior frontal gyrus and middle frontal gyrus), left superior frontal gyrus, left precuneus, right postcentral gyrus, right middle occipital gyrus, right inferior temporal gyrus, left superior temporal gyrus and middle temporal gyrus are involved in the development of Internet addiction (Liu et al. 2010). This study was limited by studying college students and a low N of nineteen.

Along with fMRI studies, EEG studies were conducted that found that those subjects classified as Internet addicts had lower brain scan activation on a game playing procedure than the normal group. This effect was different for the event related potential components N2 and for the P3 amplitudes. They had to engage in more cognitive endeavors to complete the inhibition task in the late stage (Dong et al. 2010). The IAD students also showed less efficiency in information processing and lower impulse control than their normal peers but not behaviorally (only in the EEG signal).

Studies investigating brain gray matter density (GMD) demonstrated changes in adolescents with Internet addiction using voxel-based morphometry (VBM) analysis on high-resolution T1-weighted structural magnetic resonance images. Compared with healthy controls, Internet-addicted adolescents had lower GMD in the left anterior cingulate cortex, left posterior cingulate cortex, left insula, and left lingual gyrus (Zhou et al. 2011). Again, this study used a small sample and only college students. More meaningful data would be found with a larger and more diverse subject pool. One of the most promising new biological research interests is in the area of genetic markers. The researchers from the University of Bonn and the Central Institute of Mental Health in Mannheim compared the genetic makeup of 132 problematic Internet users with that of 132 age—and sex-matched healthy control individuals (Montag et al. 2012). Results found that the 132 problem Internet users showed higher elevations of the CC genotype of rs1044396 (genetic variation of the nicotinic acetylcholine receptor gene) compared to controls. These neurotransmitters play a significant role in activating the brain's reward system.

Han et al. (2007) examined 79 adolescent male excessive Internet gamers and 75 age- and gender-matched healthy comparison adolescents. They used the reward-dependence (RD) scale in Cloninger's Temperament and Character Inventory and the frequencies of two dopamine polymorphisms: the DRD2/ANKK1 Taq Ia and COMT Val158Met polymorphisms. Their study found the excessive gamer group had significantly higher RD scores than controls. Within the EIGP group, the presence of the Taq1A1 allele correlated with higher RD scores and an increased prevalence of the DRD2 Taq1A1 and low activity COMT alleles. Lee et al. (2008) examined 91 male adolescents with excessive Internet use and 75 healthy comparison subjects. Between group comparisons were made on genetic polymorphisms of the serotonin transporter gene and with respect to harm avoidance (HA) of Cloninger's Temperament Character Inventory. Results found that the excessive Internet user group had higher prevalence of SS-genotypes, higher harm avoidance, and depression scores suggesting that excessive users may have genetic and personality traits similar to depressed patients.

1.3 Treatment Approaches

Use of the Internet is legitimate in business and home practice such as in electronic correspondence to vendors or electronic banking. Therefore, traditional abstinence models are not practical interventions when they prescribe banned Internet use in most cases. The focus of treatment consists of moderated Internet use. While moderated Internet use is the primary goal of treatment, abstinence of problematic applications is often necessary. Specific applications such as a particular game, a particular gambling site, or a particular sex site will trigger net-binges. Abstinence of the 'trigger' application is essential to help the client recover from the problematic application(s) while retaining controlled use over legitimate business Internet use.

Treatment includes a variety of interventions and a mix of psychotherapy theories to treat the behavior and address underlying psychosocial issues that are often co-existent with this addiction (e.g., social phobia, mood disorders, sleep disorders, marital dissatisfaction, or job burnout). To help clients abstain from problematic online applications, recovery interventions apply structured, measurable, and systematic techniques. The most commonly discussed therapies are Motivational Interviewing, Cognitive-Behavioral Therapy (CBT), and retreat or inpatient care.

1.3.1 Motivational Interviewing

The concept of motivational interviewing evolved from experience in the treatment of problem drinkers, and was first described by Miller (1983). These fundamental concepts and approaches were later elaborated by Miller and Rollnick (1991) in a more detailed description of clinical procedures. Motivational interviewing is a goal-directed style of counseling for eliciting behavior change by helping clients to explore and resolve ambivalence. Motivational interviewing involves asking open-ended questions, giving affirmations, and reflective listening.

Motivational interviewing is intended to confront the client in a constructive manner to evoke change, or using external contingencies such as the potential loss of a job or relationship, to mobilize a client's values and goals to stimulate behavior change. Clients dealing with addiction or substance abuse problems often feel ambivalent about quitting, even after they admit they have a problem. They fear the loss of the Internet, they fear what life might be like if they were unable to chat with online friends, engage in online activities, and use the Internet as a form of psychological escape. Motivational interview helps clients confront their ambivalence.

Typical interview questions may include: How many hours per week do you currently spend online (for non-essential use)? What applications do you use on the Internet (specific sites/groups/games visited)? How would you rank order each application from best to least important? (1 = first, 2 = second, 3 = third, etc.)? What do you like best about each application? What do you like the least? How has the Internet changed your life? How do you feel when you log offline? What problems or consequences have stemmed from your Internet use? (If this answer is difficult for the client to describe, have the client keep a log near the computer in order to document such behaviors for the next week's session)? Have others complained about how much time you spend online?

The answers to these questions create a clearer clinical profile of the client. The therapist can determine the types of applications that are most problematic for the client (i.e., chat rooms, online gaming, online pornography, etc.). The length of Internet use, the consequences of the behavior, a history of prior treatment attempts, and outcomes for any treatment attempts are also assessed. This helps clients begin the process of examining how the Internet impacts their lives.

It is helpful for the client to gain a sense of responsibility for his or her behavior. By allowing the client to resolve their ambivalence in a manner that gently pushes them, helps the client to be more inclined to acknowledge the consequences of their excessive online use and engage in treatment. Generally, the style is quiet and eliciting rather than aggressive, confrontational, or argumentative. For the therapists accustomed to confronting and giving advice, motivational interviewing can appear to be a hopelessly slow and passive process. The proof is in the outcome. More aggressive strategies, sometimes guided by a desire to "confront client denial," easily slip into pushing clients to make changes for which they are not ready.

Helping the client explore how he or she feels just before going online will help pinpoint the types of emotions being covered by the behavior (or how the client is using the Internet to cope or escape from problems). Answers may include issues such as a fight with a spouse, depressed mood, stress at a job, or a poor grade in school. Motivational interviewing should explore how these feelings diminish when online, looking for how the client rationalizes or justifies using the Internet (e.g., chatting makes me forget about the fight with my husband, looking at online porn makes me feel less depressed, gambling online makes me feel less stressed at work, killing other players in an online game makes me to feel better about my poor grade at school). Motivational interviewing is also meant to help the client recognize consequences stemming from excessive or compulsive use. Problems may consist of issues like my spouse becomes angrier, my feelings return when I turn off the computer, my job still stinks, I will lose my scholarship if I don't get my grades up. The therapeutic relationship is more like a partnership or companionship than expert/recipient roles to examine and resolve ambivalence. The operational assumption in motivational interviewing is that ambivalence is the principal obstacle to be overcome in triggering change. Overall, the specific strategies are designed to elicit, clarify, and resolve ambivalence in a client-centered and respectful therapeutic manner.

1.3.2 Cognitive-Behavior Therapy

Researchers have considered Internet addiction as a new impulse control disorder and have suggested using cognitive behavioral therapy (CBT) to treat the condition. However, given the daily dependency our society has on the Internet and technology in general unlike other impulse control issues, a specialized kind of CBT called treatment Cognitive-Behavioral Therapy for Internet Addiction (CBT-IA; Young 2011) was developed. In treating Internet addiction, abstinence recovery models are not practical as computers have become such a salient part of our daily lives. Research has found cognitive distortions are most associated with Internet addiction (e.g., Caplan 2002; Davis 2001; LaRose et al. 2001; Young 2007). CBT-IA was developed to address these cognitions. CBT-IA is a three phase approach that initially involves behavior modification to control Internet use, cognitive restructuring to challenge and modify cognitive distortions, and harm reduction therapy to address co-morbid issues.

Behavior therapy is used in the first phase of treatment to examine both computer behavior and non-computer behavior. Computer behavior deals with actual online usage, with a primary goal of abstinence from problematic applications, while retaining controlled use of the computer for legitimate purposes. Internet addicts feel a sense of displacement when online and were unable to manage central aspects of their lives due to their growing preoccupation with online use (Young 2004). They start to miss important deadlines at work, spend less time with their family, and slowly withdraw from their normal routines. They neglect social connections with their friends, coworkers, and with their communities, and, ultimately, their lives become unmanageable because of the Internet. They become consumed with their Internet activities, preferring online games, chatting with online friends, or gambling over the Internet, and ignoring family and friends in exchange for solitary time in front of the computer (Leung 2007). Managing their time online and offline is an initial goal of CBT-IA (Young 2011).

In the second phase, the rationalizations that justify excessive online use are identified, challenged, and modified. These cognitions serve as triggers for addictive behavior. For instance, Internet addicts often ruminate about their self-worth in the real world and form extreme self-concepts favoring the online self (e.g., "I am worthless offline, but in the online world I am someone important"). A gamer creates an avatar (an online game character) who achieves greater levels of success in the game so he thinks the real world is less desirable or fears that he is not as important or interesting offline. A woman who feels inadequate with meeting men offline creates an online persona where she is popular with the men online. CBT-IA uses cognitive restructuring to break this pattern. Cognitive restructuring puts the client's thoughts "under the microscope" by challenging him or her and re-scripting the negative thinking that lies behind him or her. In doing so, CBT-IA can help clients understand that they are using the Internet to avoid situations or feelings. Our moods are driven by what we tell ourselves, and this is usually based on our interpretations of our environment. Cognitive restructuring helps clients re-evaluate how rational and valid these interpretations are and find ways of achieving those same feelings offline.

Harm Reduction Therapy (HRT; Marlatt et al. 2001) is used in the third and final phase of treatment for continued recovery and relapse prevention. HRT can be used to identify and treat psychiatric issues co-existing with compulsive Internet use and/or social issues in immediate family and/or marital relationships. HRT addresses any co-existing factors associated with the development of Internet addiction. These factors can include personal, situational, social, psychiatric, or occupational issues. Often, addicts falsely assume that just stopping the behavior is enough to say, "I am recovered." Full recovery is more than simply refraining from the Internet. According to CBT-IA, complete recovery means resolving the underlying issues associated with the addiction; otherwise, relapse is likely to occur.

1.3.3 Inpatient Care and Retreat Centers

In the US, ReStart at www.netaddictionrecovery.com is one of the retreat centers specializing in problematic Internet use, video game, and technology use. They have a multidisciplinary team that works with clients in a 45-day residential care program through individualized assessments, treatment of co-occurring mental health concerns, group counseling and psychotherapy, life skills, mentoring and vocational coaching, 12-step meetings and spiritual recovery. Participants stay at Heavensfield Retreat Center in rural Fall City, Washington in the Pacific Northwest.

In China, the country has led much of the research in the Internet addiction field and established the first Internet Addiction Center in 2006, a military-run boot camp in Beijing (Jiang 2009). Tao Ran, director of the treatment center and a colonel in the People's Liberation Army (PLA), helped come up with a strict definition of Internet addiction last fall: consecutive usage of the Web for 6 h a day for three straight months is addiction. Surprisingly, almost 30 % of Chinese match this definition and more clinics have emerged. Life in the treatment camp is defined by strict, semi-military disciplines. Patients get up at 6:30 a.m. and go to bed at 9:30 p.m. Their daily schedule includes military drills, therapy sessions, reading and sports. While rigorous and controversial, outcome studies do not exist to show the efficacy of the camps.

In Korea, the most wired nation on earth has devoted the most resources to Internet addiction recovery. Perhaps no other country has so fully embraced the Internet. 90 % of homes connect to cheap, high-speed broadband, online gaming is a professional sport, and social life for the young revolves around the "PC bang," dim Internet parlors that sit on practically every street corner. But such ready access to the Web has come at a price as legions of obsessed users find that they cannot tear themselves away from their computer screens.

To address the problem, the government has built a network of 140 Internet-addiction counseling centers, in addition to treatment programs at almost 100 hospitals and the Internet Rescue camp, a forested area about an hour south of Seoul, was created to treat the most severe cases. The camp is entirely paid for by the government, making it tuition-free (Sang-Hun 2010).

Korea has been the most progressive of any country in its efforts to prevent and treat Internet addiction. To counter what is perceived as an epidemic, the government introduced a so-called "Shutdown Law", which blocks gamers under 16 from playing between midnight and 6 a.m. But its effect has been limited as teens circumvent the restrictions by using their parents' accounts (Kim and Shin 2013). The law has been semi-effective and a new trend has been suggested to enter these young people into the Riding Healing Center, a therapy organization that uses horse-riding to cure emotional and behavioral disorders, which it believes are an underlying cause of Internet addiction. The Korean Riding Association has two therapy centers and about 50 people a day go through its programs to treat a range of issues such as depression, attention hyperactivity deficit disorder (ADHD) and Internet addiction. While no outcome studies exist to their effectiveness, the association plans to build 30 more centers across South Korea, which has a population of 50 million, by 2022 to meet the rising demand for its therapy.

1.4 Final Thoughts

This chapter presents a brief overview of the studies that have been conducted on Internet addiction. Since the earlier studies in 1996, the field has grown dramatically. This chapter reviews the current diagnosis involved in detecting Internet addiction and the general set of risk factors associated with the condition. As this book focuses on the neuroscience of Internet addiction, it shows that the disorder has gained extensive credibility over the years and that new research such as this offer great opportunity to understand the underpinnings of the problem.

While not everyone who uses the Internet becomes addicted, findings show that Internet addiction is a global problem that transcends culture, race, age, and gender. As children and teenagers go online at younger ages, new risks are created and we currently know little about the lasting impact of online technologies on brain development. We may inadvertently be endangering children without realizing this impact by introducing technology at younger ages.

The research contained in this book will enable us to fully realize the potential neuroscientific impact of new technologies. This research helps us learn new diagnostic techniques for early detection. Neuroscientific studies on Internet addiction enable new psychopharmacological treatment of Internet Addiction as we learn more about the biological basis of the condition. This research helps guide the mental health field in developing new therapeutic interventions in the treatment of Internet Addiction and government agencies to develop comprehensive policies to assist in the prevention and education of Internet addiction.

The field is still in its infancy. As we search for a clearer understanding of the behavioral and the neurologic factors influencing the condition of addictive or compulsive use of the Internet. We have documented the ramifications of the disorder on social, personal, and occupational functioning. With continued work in the field, studies in structural brain imagining, functional MRIs, and molecular genetics not only assist in advancing Internet addiction research but it helps in understanding the similarities and differences among addictive syndromes overall.

References

American Psychiatric Association (1994) Diagnostic and statistical manual of mental disorders (DSM). APA, Washington D.C

Atwood JD, Schwartz L (2002) Cyber-sex: the new affair treatment considerations. J Couple Relat Ther 1:37–56

Beard KW, Wolf EM (2001) Modification in the proposed diagnostic criteria for Internet addiction. CyberPsychology Behav 4:377–383

Block JJ (2008) Issues for DSM-V: internet addiction. Am J Psychiatry 165:306–307

Caplan SE (2002) Problematic Internet use and psychosocial well-being: development of a theory-based cognitive-behavioral measurement instrument. Comput Hum Behav 18:553–575

Davis RA (2001) A cognitive behavioral model of pathological internet use. Comput Hum Behav 17:187–195

Dong G, Lu Q, Zhou H, Zhao X (2010) Impulse inhibition in people with internet addiction disorder: electrophysiological evidence from a Go/NoGo study. Neurosci Lett 485:138–142

Dowling NA, Quirk KL (2009) Screening for Internet dependence: do the proposed diagnostic criteria differentiate normal from dependent internet use? CyberPsychology Behav 12:21–27

Ebeling-Witte S, Frank ML, Lester D (2007) Shyness, Internet use, and personality. CyberPsychology Behav 10:713–716

Ferraro G, Caci B, D'Amico A, Di Blasi M (2007) Internet addiction disorder: an Italian study. CyberPsychology Behav 10:170–175

Ghassemzadeh L, Shahraray M, Moradi A (2008) Prevalence of Internet addiction and comparison of internet addicts and non-addicts in Iranian High Schools. Cyberpsychology Behav 11:731–733

Han DH, Lee YS, Yang KC et al (2007) Dopamine genes and reward dependence in adolescents with excessive internet video game play. J Addict Med 1:133–138

Hardie E, Tee MY (2007) Excessive internet use: the role of personality, loneliness and social support networks in internet addiction. Aust J Emerg Technol Soc 5:34–47

Hur MH (2006) Internet addiction in Korean teenagers. CyberPsychology Behav 9:14–525

Jiang J (2009) Inside China's fight against internet addiction. In: Time. Available via: http://www.time.com/time/world/article/0,8599,1874380,00.html

Johansson A, Götestam KG (2004) Internet addiction: characteristics of a questionnaire and prevalence in Norwegian youth (12–18 years). Scand J Psychol 45:223–229

Kaltiala-Heino R, Lintonen T, Rimpelä A (2004) Internet addiction? Potentially problematic use of the internet in a population of 12 to 18 year old adolescents. Addict Res Theory 12:89–96

Khazaal Y, Billieux J, Thorens G et al (2008) French validation of the internet addiction test. CyberPsychology Behav 11:703–706

Kim D, Shin E (2013) Horses to the rescue of Korea's internet-addicted teens. In: Reuters. Available via: http://www.reuters.com/article/2013/01/11/us-korea-internet-horses-idUSBRE90A14G20130111. Accessed on 3 Jan 2015

Korkeila J, Kaarlas S, Jääskeläinen M et al (2010) Attached to the web—harmful use of the internet and its correlates. Eur Psychiatry 25:236–241

Lam LT, Peng Z, Mai J, Jing J (2009) Factors associated with internet addiction among adolescents. Cyberpsychology Behav 12:551–555

LaRose R, Mastro D, Eastin MS (2001) Understanding internet usage. A social-cognitive approach to uses and gratifications. Soc Sci Comput Rev 19:395–413

Lavin M, Marvin K, McLarney A et al (1999) Sensation seeking and collegiate vulnerability to internet dependence. CyberPsychology Behav 2:425–430

Lee YS, Han DH, Yang KC et al (2008) Depression like characteristics of 5HTTLPR polymorphism and temperament in excessive internet users. J Affect Disord 109:165–169

Leung L (2007) Stressful life events, motives for Internet use, and social support among digital kids. CyberPsychology Behav 10:204–214

Liu J, Gao XP, Osunde I et al (2010) Increased regional homogeneity in internet addiction disorder a resting state functional magnetic resonance imaging study. China Med J 123:1904–1908

Marlatt GA, Blumne AW, Parks GA (2001) Integrating harm reduction therapy and traditional substance abuse treatment. J Psychoactive Drugs 33:13–21

Miller WR (1983) Motivational interviewing with problem drinkers. Behav Psychother 11:147–172

Miller WR, Rollnick S (1991) Motivational interviewing: preparing people to change addictive behavior. Guilford Press, New York

Montag C, Flierl M, Markett S et al (2011) Internet addiction and personality in first-person-shooter video gamers. J Media Psychol Theor Methods Appl 23:163–173

Montag C, Kirsch P, Sauer C, Markett S, Reuter M (2012) The role of the CHRNA4 gene in internet addiction: a case-control study. J Addict Med 6:191–195

Morahan-Martin J (1999) The relationship between loneliness and internet use and abuse. CyberPsychology Behav 2:431–439

Pawlikowski M, Brand M (2011) Excessive Internet gaming and decision making: do excessive World of Warcraft players have problems in decision making under risky conditions? Psychiatry Res 188:428–433

Ryu EJ, Choi KS, Seo JS, Nam BW (2004) The relationships of internet addiction, depression, and suicidal ideation in adolescents. J Korean Acad Nurs 34:102–110

Sang-Hun C (2010) South Korea expands aid for internet addiction. New York Times. Available via: http://www.nytimes.com/2010/05/29/world/asia/29game.html?_r=0. Accessed ???

Sariyska R, Reuter M, Bey K et al (2014) Self-esteem, personality and Internet addiction: a cross-cultural comparison study. Personality Individ Differ 61–62:28–33
Shapiro NA, Goldsmith TD, Keck PE et al (2000) Psychiatric evaluation of individuals with problematic internet use. J Affect Disord 57:267–272
Siomos KE, Dafouli ED, Braimiotis DA et al (2008) Internet addiction among Greek adolescent students. Cyberpsychology Behav 11:653–657
Suhail K, Bargees Z (2006) Effects of excessive internet use on undergraduate students in Pakistan. CyberPsychology Behav 9:297–307
Tao R, Huang X, Wang J et al (2010) Proposed diagnostic criteria for internet addiction. Addiction 105:556–564
Whitty M (2005) The realness of cybercheating. Soc Sci Comput Rev 23:57–67
Widyanto L, McMurren M (2004) The psychometric properties of the internet addiction test. CyberPsychology Behav 7:445–453
Yen JY, Ko CH, Yen CF et al (2007) The comorbid psychiatric symptoms of internet addiction: attention deficit and hyperactivity disorder (ADHD), depression, social phobia, and hostility. J Adolesc Health 41:93–98
Yen JY, Ko CH, Yen CF et al (2008) Psychiatric symptoms in adolescents with internet addiction: comparison with substance use. Psychiatry Clin Neurosci 62:9–16
Young KS (1998) Internet addiction: the emergence of a new clinical disorder. CyberPsychology Behav 1:237–244
Young KS (2004) Internet addiction: the consequences of a new clinical phenomena. In: Doyle K (ed) American behavioral scientist: psychology and the new media. Sage, Thousand Oaks, pp 1–14
Young KS (2007) Cognitive-behavioral therapy with internet addicts: treatment outcomes and implications. CyberPsychology Behav 10:671–679
Young KS (2011) CBT-IA: the first treatment model to address internet addiction. J Cognitive Therapy 25:304–312
Zhou Y, Lin FC, Du YS et al (2011) Gray matter abnormalities in internet addiction: a voxel-based morphometry study. Eur J Radiol 79:92–95

Part II
Neuroscientific Approaches to Internet Addiction

Chapter 2
Structural Brain Imaging and Internet Addiction

Fuchun Lin and Hao Lei

Abstract In recent years, neuroimaging techniques have increasingly been used to study Internet addiction disorder (IAD), with the aim of identifying functional and structural changes in the brain, which may constitute the neurological/psychiatric causes of IAD. This chapter reviews current neuroimaging findings concerning brain structural changes associated with IAD. To aid readers in understanding these findings, the commonly used structural imaging methodologies—primarily, magnetic resonance imaging (MRI)—are also outlined. The literature review clearly demonstrates that IAD is associated with neuroanatomical changes involving prefrontal cortex, thalamus, and other brain regions. At least some of these changes appear to correlate with behavioral assessments of IAD. More importantly, these data suggest that the pattern of IAD-related structural differences in the brain resemble, to some extent, those changes observed in substance addiction.

2.1 Introduction

Internet addiction disorder (IAD) was originally proposed as a mental disorder in a satirical hoax by Ivan Goldberg in 1995. It commonly refers to one's inability to control his or her urge to be on-line, resulting in uncontrolled use of the Internet and adverse consequences in life, such as marked distress, impaired social interaction and loss of educational/occupational interests (Aboujaoude 2010; Douglas et al. 2008; Kuss et al. 2013). IAD, or pathological Internet use, may be caused by a spectrum of on-line activities including gaming, shopping, gambling, viewing pornography, and social networking. Clinical studies have demonstrated that subjects with uncontrolled use of the Internet, not only share core symptoms with substance

F. Lin · H. Lei (✉)
National Center for Magnetic Resonance in Wuhan, State Key Laboratory of Magnetic Resonance and Atomic and Molecular Physics, Wuhan Institute of Physics and Mathematics, Chinese Academy of Sciences, Wuhan 430071, People's Republic of China
e-mail: leihao@wipm.ac.cn

C. Montag and M. Reuter (eds.), *Internet Addiction*, Studies in Neuroscience, Psychology and Behavioral Economics, DOI 10.1007/978-3-319-07242-5_2

addiction such as tolerance, withdrawal symptoms and relapse (Beard and Wolf 2001; Young 1998), but also frequently have psychiatric co-morbidity, including attention deficit/hyperactivity disorder, anxiety disorders, sleep disorders, and obsessive-compulsiveness (Bernardi and Pallanti 2009; Ko et al. 2012; Yen et al. 2007).

Although the concept of IAD is well received by the general public and has attracted extensive popular media coverage, controversy exists among the scientific community regarding whether IAD constitutes a stand-alone illness (Chakraborty et al. 2010; Morahan-Martin 2005). Currently, IAD is not officially recognized as a psychiatric disorder in most parts of the world. In the newly released Diagnostic and Statistical Manual of Mental Disorders Edition V (DSM-V), Internet gaming disorder, which constitutes a major subtype of IAD, is listed as one of the "conditions for further study" (http://www.dsm5.org/Pages/Default.aspx).

With reference to the criteria defining pathological gambling and substance addiction, psychometric tools have been constructed for IAD assessment, among which the Young's Internet addiction scale (YIAS) (Young 1996) and Young's diagnostic questionnaire for Internet addiction (YDQ) (Young 1998), both developed by Dr. Kimberly Young, are the most widely used. Although discrepancy and controversy still exist around such criteria, they nonetheless provide a common ground for communication and research on IAD, and have been widely used in practice.

As with many other psychiatric disorders, the fiercest debates swirling around IAD concern the problem of defining the condition scientifically. Entering the era of DSM-V, more and more neurologists, psychiatrists and researchers would agree that defining a psychiatric/mental disorder, such as addiction, solely based on symptomatology (or psychometric assessment) may not be sufficient. More objective biomarkers, such as genetic risk factor, biochemical profile and functional/structural changes of the brain, need to be uncovered to help achieve better understanding, diagnosis and treatment. Undoubtedly, neuroimaging can play a crucial role in this regard.

Because of their noninvasiveness and capability of providing functional/structural information on the brain in high spatial resolution, neuroimaging approaches, especially magnetic resonance imaging (MRI), have been increasingly used over the last two decades to study the neural mechanisms underlying psychiatric disorders. Through neuroimaging research, many psychiatric disorders originally thought to have no clear anatomical pathology are now known to be associated with functional/structural abnormalities of the brain at the neural circuit/network level. For example, subjects addicted to substances were consistently shown to have prominent functional as well as structural changes in the prefrontal cortex (PFC), and such PFC abnormalities are known to play crucial roles in the development of craving, compulsive use and relapse (Goldstein and Volkow 2011).

IAD is believed by some to be a form of so-called behavioral addiction, which is expected to share similar neural mechanisms, at least in part, with substance addiction. However, there are also researchers who disagree with this concept; skeptical about whether non-drug stimuli, such as repetitive, high-frequency and highly-rewarding behaviors/experiences, could be potent enough to generate neuroadapation similar to that found in substance addiction (Holden 2001). One way to settle this disagreement and lead to a better understanding/definition of IAD is to see whether the functional/structural abnormalities known to be associated with substance addiction, as revealed by neuroimaging approaches, are also present in subjects with IAD (as defined by psychometric assessments). In fact, an increasing number of such studies have been done in the past few years. In this chapter, we shall focus on neuroimaging findings on the brain structural abnormalities associated with IAD. Literature results are reviewed, and the implications of the findings are discussed.

2.2 Methodologies for Assessing Structural Changes of the Brain

2.2.1 Three-Dimensional Anatomical MRI

Among the existing neuroimaging approaches, MRI is probably the most powerful and widely used for assessing structural changes of the brain. Three-dimensional (3D) T_1-weighted imaging (T_1WI) is the most commonly used technique for anatomical MRI, because it is fast in terms of acquisition speed, and is capable of providing high-resolution images with clear contrast among grey matter (GM), white matter (WM) and cerebrospinal fluid (CSF). Moreover, 3D acquisition enables reconstruction of brain slices in any arbitrary orientation. With the state-of-art technology, a 3D-T_1WI dataset covering the whole human brain can be acquired in less than 10 min on the 3 Tesla clinical scanners with an isotropic spatial resolution of 0.5 mm. With different image processing methods, volumetric and morphormetric measures could be derived from the whole-brain 3D-T_1WI dataset, and such measures are often used to assess structural changes of the brain.

2.2.1.1 Volumetric Analysis

First, the 3D-T_1WI dataset can be used for quantitative volumetric analysis of the whole brain as well as any given brain structure of interest. To measure the volume of the whole brain, the non-brain voxels on the images are removed either manually or automatically using special algorithms. The number of brain voxels can then be counted and used to derive the volume of the whole brain. To derive the volume of a given brain structure, a region-of-interest (ROI) representing the structure under concern is delineated, usually manually and with reference to the landmarks on the images, and the number of voxels in the ROI can then be counted and used to derive the volume. An atlas is often needed to guide the delineation of the ROI. For example, Makris et al. (2008b) used this method to found out that long-term alcohol users had significantly decreased reward-network (i.e., dorsolateral PFC (dlPFC) and insula) volume than normal controls. However, this method can be laborious, and the results are susceptible to objective bias and cross-subject variations in how the ROI is delineated.

2.2.1.2 Voxel-Based Morphometry

Voxel-based morphometry (VBM) is an unbiased objective technique developed to characterize subtle structural changes in the whole brain, without the need of any *a prior* knowledge (Ashburner and Friston 2000). The aim of VBM is to identify differences in the local composition of GM and WM at the group level. VBM involves spatially normalizing the anatomical imaging data from individual subjects into the same stereotactic space, segmenting the individual normalized images into GM/WM/CSF compartments, smoothing the segmented images spatially, and performing voxel-wise statistical analyses to localize significant inter-group differences. The output of VBM is a statistical parametric map showing brain regions where GM/WM composition differs significantly at the group level (Ashburner and Friston 2000).

GM (or WM) density and GM (or WM) volume are two frequently used measures of tissue composition in VBM analysis. Although the two are related to each other, they differ conceptually. Within a voxel on the spatially normalized images or an ROI, GM/WM density means the relative concentration of GM/WM tissue (i.e. the proportion of GM/WM to all tissue types), while GM/WM volume means the absolute GM/WM volume. Comparing GM/WM volume within the framework of VBM involves multiplying the spatially normalized GM/WM density by its Jacobian determinants derived from deformation flow information (Mechelli et al. 2005).

VBM analysis has been widely used in neuroimaging studies of addiction (Barros-Loscertales et al. 2011; Liao et al. 2012; Liu et al. 2009; Makris et al. 2008b; Schwartz et al. 2010). Such studies consistently show that subjects who are dependent on stimulant drugs have significantly reduced GM volume in the PFC (Ersche et al. 2013).

2.2.1.3 Cortical Thickness Measurement

In addition to volumetric measures and regional composition of GM/WM, 3D-T_1WI dataset can also be used to derive cortical thickness, a computational neuroanatomy measure defined as the distances between the pial surface (i.e., surface between the cortical GM and CSF) and the interface separating the cortical GM and WM underneath (MacDonald et al. 2000). Measuring cortical thickness involves segmenting the anatomical images into GM/WM/CSF compartments for each individual subject, reconstructing the individual GM/WM surfaces and pial surface, computing individual cortical thickness, registering the surface-based coordinate system of each individual subject into the same stereotactic space, spatial smoothing, and performing voxel-wise statistics to detect morphometric variations at the group level.

Figure 2.1 explains graphically the steps to measure cortical thickness from the 3D anatomical image data. Cortical thickness is thought to be related to the size, density and arrangement of cortical cells (MacDonald et al. 2000), and has been shown to change only minimally with brain size and sex (Sowell et al. 2007). Typical cortical thickness values in adult humans are between 1.5 and 3 mm (Salat et al. 2004). Cortical thickness has been used to investigate structural changes of the cortices associated with neurodevelopment and brain diseases. During aging, a decrease (also known as cortical thinning) on the order of about 10 μm per year has been observed (Salat et al. 2004).

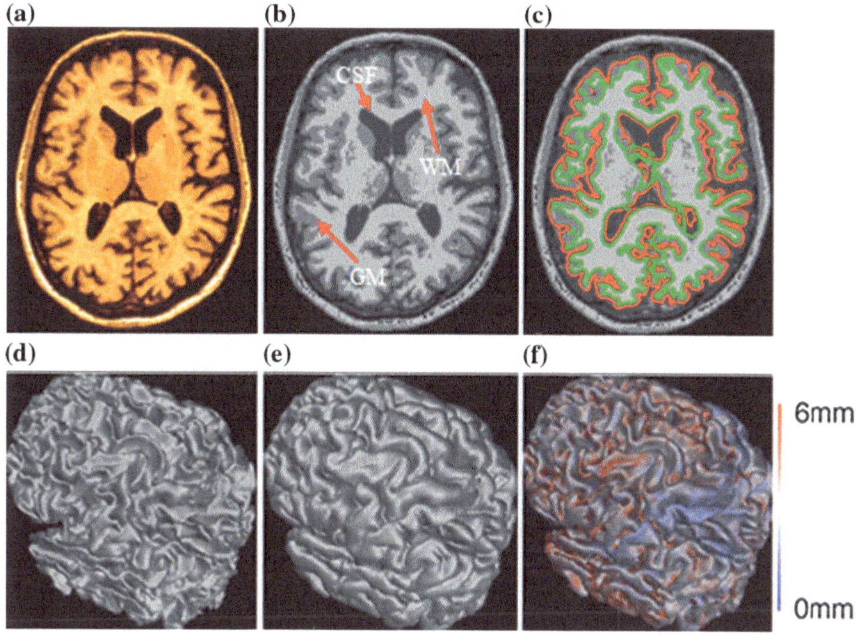

Fig. 2.1 Segmentation and cortical thickness analysis of anatomical images. The raw anatomical images were first corrected for signal intensity nonuniformity and registered into a reference stereotaxic space (**a**), and then segmented into grey matter (*GM*), white matter (*WM*) and cerebrospinal fluid (*CSF*) compartments (**b**). The inner (*green lines* in **c**) and outer (*red lines* in **c**) GM surfaces can then be extracted and fitted into three dimensional maps using deformable models. Panel **d** shows the resultant inner GM surface, and panel **e** shows the outer GM surface. At a given coordinate in the reference stereotaxic space, cortical thickness is defined as the distance between these two surfaces (**f**). This figure is adapted from a figure from the paper by Lerch et al. (2005) with permission

Cocaine-dependent subjects are known to have lower cortical thicknesses in brain regions involved in executive regulation of reward and attention (Makris et al. 2008a).

2.2.2 Diffusion Tensor Imaging

Diffusion tensor imaging (DTI) is a noninvasive MRI technique that measures the diffusion properties of water molecules *in vivo* (Basser et al. 1994a, b; Le Bihan 2003; Pierpaoli et al. 1996). The diffusion of water molecules in a homogenous compartment, such as CSF, is isotropic, and can be characterized simply by a single diffusion coefficient. However, in biological tissues, the diffusion of water molecules is subject to restriction imposed by the microstructural organization of the tissue (e.g., membranes and other biological barriers). For instance, in the WM fibers, the water molecules would diffuse more quickly along the fibers than perpendicular to the fibers. As a result, more complicated models need to be used to characterize the anisotropic diffusion properties of water molecules in the biological tissues.

In DTI, the diffusion behaviors of water molecules are modeled by a zero mean Gaussian distribution, which is fully represented by a second order diffusion tensor (Basser et al. 1994a, b; Pierpaoli et al. 1996). After measuring the diffusion tensor experimentally, parameterized diffusion indices, such as fractional anisotropy (FA), can be computed (Basser and Pierpaoli 1996). FA is a scalar value between zero and one that describes the degree of anisotropy of a diffusion process. A value of zero means that diffusion is isotropic (i.e., it is unrestricted or equally restricted in all directions). A value of one means that diffusion occurs only along one axis, and is fully restricted along all the other directions.

Measuring the diffusion indices of water molecules along different directions and the overall anisotropy with DTI may provide important information on the microstructural organization of the underlying tissue (Le Bihan 2003). For example, the FA value of a WM tract is thought to be closely related to fiber density, axonal diameter and myelination, thus often being used as a surrogate for the assessing the microstructural integrity of WM. It has been demonstrated that diffusion indices obtained from DTI can be used to detect tissue microstructural changes that might not be visible with the conventional MRI techniques (Basser et al. 1994a, b; Pierpaoli et al. 1996). Nowadays, DTI has become a widely used tool for revealing the tissue abnormalities associated with neurological/psychiatric diseases. Figure 2.2 shows representative DTI data that are commonly used in the studies on disease-related brain structural changes.

2.2.2.1 Voxel-Based Analysis

Voxel-based analysis (VBA) is an observer-independent voxel-wise analysis method for diffusion indices derived from the DTI data, which can circumvent the problems associated with the more traditional ROI analysis (Jones et al. 2005). The aim of VBA is to assess regional alterations of diffusion indices between groups. VBA includes spatially normalizing the maps of diffusion indices from individual

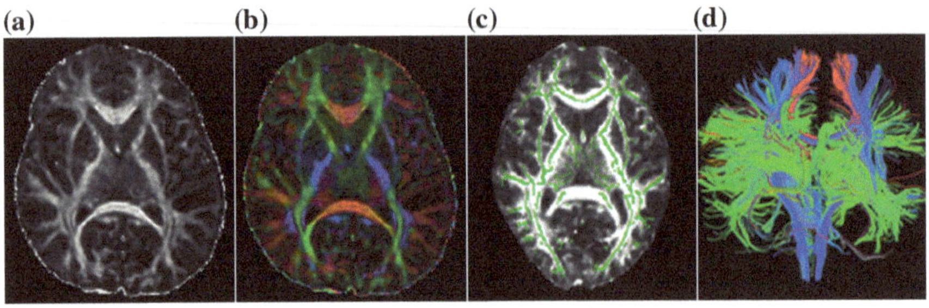

Fig. 2.2 Diffusion tensor imaging (*DTI*) and data analysis. With images acquired with diffusion-weighted gradients applied along different directions and a tensor model, fractional anisotropy (*FA*) maps (**a**), corresponding FA-weighted color directional diffusion maps (**b**), FA skeleton maps (**c**) can be calculated. Whole-brain tractography (**d**) can be performed. The data in **a**–**d** are from the same normal subject

subjects into a standard stereotactic space, smoothing the normalized maps, and performing voxel-by-voxel statistical comparisons to determine significant inter-group differences. With VBA, the whole brain is tested for control-patient differences without any *a priori* hypothesis on where the abnormalities should be. The output of VBA is a statistical parametric map showing brain regions where diffusion indices differ significantly at the group level.

2.2.2.2 Tract-Based Spatial Statistics

Tract-based Spatial Statistics (TBSS) is another observer-independent voxel-wise method for analyzing whole-brain DTI data (Smith et al. 2006). TBSS involves co-registering the FA maps from all individuals included in the study, to a standard stereotactic space, averaging the co-registered individual FA maps to create a mean FA map, thinning the mean FA map to obtain a mean FA skeleton, projecting co-registered individual FA maps onto the mean FA skeleton to create a skeletonized FA map, and finally performing voxel-wise statistics across subjects on the skeletonized FA data. TBSS retains the strengths of VBA while addressing some of its drawbacks, such as the arbitrariness of the choice of spatial smoothing.

2.2.2.3 Tractography-Based Analysis

DTI data can also be used to trace WM tracts by performing tractography according to the principal directions of neighboring diffusion tensors. The main feature of DTI tractography is that it can be used to reconstruct WM pathways *in vivo*, and provide information about the shape, location and topology of fiber tracts as well as anatomical connectivity between distant brain areas (Basser et al. 2000; Conturo et al. 1999; Mori et al. 1999). Tractography is a useful tool for measuring WM deficits, and has been applied in a wide range of clinical and basic studies (Dell'Acqua and Catani 2012). In tractography-based analysis, the fiber tract under concern is first reconstructed using fiber tracking algorithms, and the diffusion indices of the tract can then be analyzed by considering the fiber tract as a 3D ROI (McIntosh et al. 2008) or by parameterizing the fiber tract (Lin et al. 2006).

2.3 Brain Structural Abnormalities Associated with IAD

Unlike the case for substance addiction, only a limited number of structural neuroimging studies have been performed on IAD so far, mostly by Chinese and Korean researchers. Table 2.1 lists all the structural neuroimging studies on IAD that can be found in the literature, and summarizes the major findings from these studies. We shall also give a brief review of these results below.

Table 2.1 A summary of the structural brain imaging studies on Internet addiction disorder (IAD) available so far

Studies	Subjects	Diagnosis criteria	Methodology	Main results
Zhou et al. (2011)	18 addicts (16M/2F), 17.2 ± 2.6 years old 15 controls (13M/2F), 17.8 ± 2.5 years old	Modified young diagnostic questionnaire for internet addiction by Beard and Wolf (MYDQ)	Scanner: 3.0T Philips Achieva Acquisition: three-dimensional (3D) T_1-weighted anatomical imaging Analysis: Voxel-based morphometry (VBM) to assess gray matter density (GMD)	Compared to the controls, the addicts showed decreased gray matter density (GMD) in: Left anterior cingulate cortex (ACC) Left posterior cingulate cortex (PCC) Left insula Left lingual gyrus
Yuan et al. (2011)	18 addicts (12M/6F), 19.4 ± 3.1 years old 18 controls (12M/6F), 19.5 ± 2.8 years old	MYDQ Self-rating anxiety scale and the self-rating depression scale Subjects pending 19.4 ± 3.1 h per day playing on-line games	Scanner: 3.0T Siemens Allegra Acquisition: 3D T_1-weighted anatomical imaging + diffusion tensor imaging (DTI) Analysis: VBM to assess gray matter volume (GMV) + tract-based spatial statistics (TBSS)	Compared to the controls, the addicts showed decreased gray matter volume (GMV) in: Bilateral orbitofrontal cortex (OFC) Bilateral supplementary motor area (SMA) Bilateral dorsolateral prefrontal cortex (dlPFC) Left rostral ACC (rACC) Cerebellum and reduced fractional anisotropy (FA) in: White matter within the right parahippocampal gyrus and increased FA in: Left posterior limb of the internal capsule (PLIC) GMVs of the right dlPFC, left rACC, and right SMA in the addicts correlated negatively with the duration of addiction FA of the left PLIC correlated positively with the duration of addiction

(continued)

Table 2.1 (continued)

Studies	Subjects	Diagnosis criteria	Methodology	Main results
Dong et al. (2012)	16 addicts (16M/0F), 22.2 ± 3.3 years old 15 control (15M/0F), 21.6 ± 2.6 years old	Young's online internet addiction test Structured psychiatric interviews (M.I.N.I.) Subjects spending > 80 % online time playing online games	Scanner: 3T Siemens Trio Acquisition: DTI Analysis: TBSS	Compared with the controls, the addicts showed increased FA in: Bilateral thalamus Left PCC Thalamic FA in the addicts correlated positively with Internet addiction severity scores
Han et al. (2012)	20 addicts (20M/0F), 20.9 ± 2.0 years old 17 professional gamers (PG) who were not addicted (17M/0F), 20.8 ± 1.5 years old 18 controls (18M/0F), 20.9 ± 2.1 years old	Young's internet addiction scale (YIAS) Structured clinical interview for DSM-IV and the beck depression inventory Subjects spending > 4 h per day/30 h per week playing on-line games	Scanner: 1.5 T Siemens Espree Acquisition: 3D T_1-weighted anatomical imaging Analysis: VBM to assess changes in GMV	Compared to the controls, the addicts showed increased GMV in: Left thalamus and decreased GMV in: Bilateral inferior temporal gyri Right middle occipital gyrus Left inferior occipital gyrus Compared to the PG who were not addicted, the addicts showed increased GMV in: Left thalamus and decreased GMV in: Left cingulate gyrus (CG) GMV of the left CG in the addicts correlated negatively with the YIAS scores and Barratt Impulsiveness Scale total scores GMV of the thalamus in the addicts correlated positively with the YIAS scores

(continued)

Table 2.1 (continued)

Studies	Subjects	Diagnosis criteria	Methodology	Main results
Lin et al. (2012)	17 addicts (15M/2F), 17.0 ± 2.5 years old 16 controls (14M/2F), 17.8 ± 2.5 years old	MYDQ Mini international neuropsychiatric interview for children and adolescents	Scanner: 3.0 T Phillips Achieva Acquisition: DTI Analysis: TBSS	Compared to the controls, the addicts showed decreased FA in: Bilateral orbito-frontal white matter Genu of corpus callosum (CC) Bilateral anterior cingulum Bilateral inferior fronto-occipital fasciculus Bilateral corona radiation Bilateral anterior limb of internal capsule External capsule (EC) Left precentral gyrus FA of left genu of CC of the addicts correlated negatively with the screen for child anxiety related emotional disorders FA of the left EC of the addicts correlated negatively with the YIAS scores
Hong et al. (2013)	15 addicts (15M/0F), 13.3 ± 2.8 years old 15 controls (15M/0F), 15.4 ± 1.2 years old	YIAS Kiddie-Schedule for Affective Disorders and Schizophrenia-present and lifetime version Subjects self-report to have experienced typical components of addiction to online gaming	Scanner: 3T Siemens Trio Acquisition: 3D T_1-weighted anatomical imaging Analysis: cortical thickness	Compared to the controls, the addicts showed decreased cortical thickness in: Right lateral OFC

(continued)

Table 2.1 (continued)

Studies	Subjects	Diagnosis criteria	Methodology	Main results
Weng et al. (2013)	17 (4M/13F), 16.3 ± 3.0 years old 17 (2M/15F), 15.5 ± 3.2 years old	MYDQ Playing online game was the primary activity when the addicts used Internet	Scanner: 3.0 T Philips Intera Acquisition: 3D T_1-weighted anatomical imaging +DTI Analysis: VBM to assess GMV + TBSS	Compared to the controls, the addicts showed decreased FA in: Genu of CC Bilateral frontal lobe white matter Right EC and reduced GMV in: Right OFC Bilateral insula Right SMA GMVs of the right OFC and bilateral insula correlated negatively with the YIAS scores FA of the right EC correlated negatively with the YIAS scores
Yuan et al. (2013)	18 addicts (12M/6F), 19.4 ± 3.1 years old 18 controls (12M/6F), 19.5 ± 2.8 years old	MYDQ	Scanner: 3.0 T Siemens Allegra Acquisition: 3D T_1-weighted anatomical imaging Analysis: cortical thickness	Compared to the controls, the addicts showed decreased cortical thickness in: Left lateral OFC Left insula Left lingual gyrus Right postcentral gyrus Right entorhinal cortex Right inferior parietal cortex and increased cortical thickness in: Left precentral cortex Left precuneus Left middle frontal cortex Left inferior temporal and middle temporal cortices Cortical thicknesses of the left precentral cortex and precuneus in addicts correlated positively with the duration of addiction Cortical thickness of the left lingual gyrus correlated negatively with the duration of addiction

2.3.1 Results from Anatomical MRI

2.3.1.1 VBM Analysis

Zhou et al. (2011) were among the first to use a neuroimging approach to assess structural abnormalities in the brain associated with IAD. They acquired 3D-T_1WI data from 18 adolescents (i.e., 17.2 ± 2.6 years) who were considered to be addicted to the Internet based on the criteria of the modified eight-item YDQ (Beard and Wolf 2001), and 15 age- and gender-matched healthy controls. VBM analysis was used to compare regional GM density (GMD) between the two groups. It was reported that the IAD group had significantly reduced GMD in the left anterior cingulate cortex (ACC), left posterior cingulate cortex (PCC), left insula and left lingual gyrus. The major online-activities of the IAD subjects were not specified in this study.

There have been three studies that investigated structural abnormalities in the brain of adolescent/young subjects (i.e., 16–21 years) who were specifically addicted to on-line games (Han et al. 2012; Weng et al. 2013; Yuan et al. 2011). YDQ or YIAS was used in these studies to screen for on-line game addiction (OGA). Additionally, it was confirmed that playing online game was the primary activity for the addicted subjects when they used the Internet (i.e., on average around 10 h of on-line game playing per day). Two studies showed similar results in that, compared to normal controls, the subjects with OGA had significantly reduced GM volume (GMV) in the orbitofrontal cortex (OFC) and supplementary motor area (SMA). OGA was also found to be associated with reduced GMV in the left rostral ACC (rACC), bilateral dlPFC and cerebellum by Yuan et al. (2011); and with reduced GMV in the bilateral insula by Weng et al. (2013). In contrast to the observations in these two studies, Han et al. (2012) reported that the subjects with OGA had reduced GMV in the bilateral inferior temporal gyri, right middle occipital gyrus, and left inferior occipital gyrus, but increased GMV in the left thalamus, compared to the normal controls. They also compared regional GMV between the subjects with OGA and professional gamers who were not addicted, and found significantly lower left cingulate gyrus GMV in the addiction group (Han et al. 2012).

2.3.1.2 Cortical Thickness Analysis

There have been two studies performed so far to assess the OGA-related abnormalities in cortical thickness. Yuan et al. (2013) showed that, compared to normal controls, subjects with OGA in late adolescence had increased cortical thickness in the left precentral cortex, precuneus, middle frontal cortex, inferior temporal and middle temporal cortices, and decreased cortical thickness in the left lateral

OFC, insula, lingual gyrus, right postcentral gyrus, entorhinal cortex, and inferior parietal cortex. Hong et al. (2013) reported decreased cortical thickness in the right lateral OFC of male adolescents who were addicted to on-line gaming.

2.3.2 Results from DTI

Yuan and colleagues were among the first to use DTI to assess WM abnormalities associated with IAD. Their results showed that, relative to normal controls, adolescent college students with OGA were associated with significantly increased FA in the left posterior limb of the internal capsule (PLIC), but reduced FA in the WM within right parahippocampal gyrus (Yuan et al. 2011). Higher FA in the bilateral thalamus and left PCC were also reported in the subjects with OGA (Dong et al. 2012).

With the same IAD and control subjects as those reported in the study by Zhou et al. (2011), Lin et al. (2012) reported that IAD is associated with reduced FA in the orbito-frontal WM, corpus callosum (CC), cingulum, inferior front-occipital fasciculus, corona radiation, anterior limb of the internal capsule (ALIC) and external capsule (EC). These findings were largely reproduced in a subsequent study conducted by Weng et al. (2013), showing that adolescents with OGA had decreased FA in the right genu of CC, bilateral frontal WM and right EC, as compared to normal controls.

2.3.3 Correlations Between Brain Structural Alterations and Behavioral Assessments

Some of the studies also assessed the correlations between brain structural alterations and behavioral assessments in Internet addicts. For example, two studies on OGA showed consistently that the GMV of left CG, right OFC and bilateral insula correlated negatively with the YIAS scores and Barratt impulsiveness scale total scores; while GMV of the left thalamus correlated positively with the YIAS scores (Han et al. 2012; Weng et al. 2013). The studies of Yuan et al. (2013, 2011) on OGA showed that the GMV in right dlPFC, left rACC and right SMA, and the cortical thickness of left lingual gyrus correlated negatively with the duration of Internet addiction. Positive correlation between the cortical thickness of the left precentral cortex and precuneus and the duration of Internet addition was also reported (Yuan et al. 2013).

DTI studies revealed that the addicted subjects had a negative correlation between FA in the EC and YIAS scores (Lin et al. 2012; Weng et al. 2013), and positive correlations between FA in the thalamus and YIAS cores (Dong et al. 2012).

A positive correlation between FA of the left PLIC and the duration of Internet addiction was also reported (Yuan et al. 2011). Additionally, Lin et al. (2012) reported a negative correlation between FA of the left genu of CC and the Screen for Child Anxiety Related Emotional Disorders scores.

2.3.4 Synopsis of Structural Abnormalities Associated with IAD/OGA

The structural neuroimaging results summarized in Sects. 2.3.1–2.3.3 consistently demonstrate that IAD and/or OGA is associated with structural abnormalities in the brain, although the exact pattern and characteristics of the abnormalities may appear to vary from study to study. The most consistent findings from the studies available so far are atrophy in the PFC (i.e., OFC, ACC and dlPFC) and insula. Almost all the studies demonstrate reduced GMD, GMV or cortical thickness in these two regions, and such changes also appear to be correlated with either the YIAS scores or the duration of Internet addiction.

Thalamus is another brain region frequently reported to show structural abnormalities in subjects with IAD/OGA. But unlike the case for PFC and insula, the findings for thalamus appeared to be less consistent. Increased thalamic GMV and FA have been reported in IAD/OGA, and the increase in thalamic FA was shown to correlate positively with the YIAS scores (Dong et al. 2012; Han et al. 2012). On the other hand, although no structural changes in the thalamus was reported in the paper by Zhou et al. (2011), a trend toward decreased GMD in the bilateral anterior thalamus was found for the subjects with IAD (Fig. 2.3).

Other brain regions found to demonstrate structural changes were mainly visual-related (i.e., occipital gyrus, inferior temporal gyrus and lingual gyrus) and sensory/motor-related (i.e., SMA, precentral/postcentral cortex and cerebrullum) areas. DTI abnormalities associated with IAD/OGA were found to be predominately located in or along the WM tracts connecting to PFC and thalamus, such as the genu of the CC, ALIC, EC and cingulum.

2.3.5 Comparisons with Brain Structural Abnormalities in Substance Addiction and Pathological Gambling

Abnormal GMD/GMV in the prefrontal regions (i.e., OFC, ACC and dlPFC), insula, and thalamus are common findings in smokers (Zhang et al. 2011), heroin-dependent individuals (Yuan et al. 2010), alcoholics (Makris et al. 2008b), opiate-dependent subjects (Lyoo et al. 2006), methamphetamine abusers (Kim et al. 2006) and cocaine-dependent subjects (Franklin et al. 2002). Impaired WM integrity in the orbito-frontal regions, CC, cingulum, ALIC, EC and corona radiation

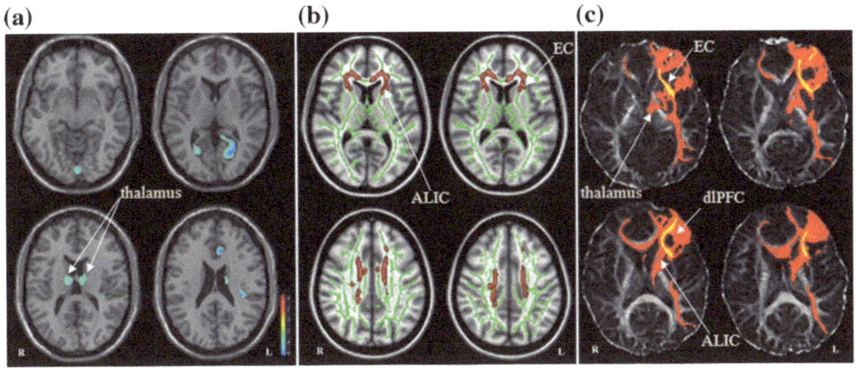

Fig. 2.3 Structural abnormalities associated with Internet addiction disorder (*IAD*) as revealed by voxel-based morphometry (*VBM*) and tract-based spatial statistics (*TBSS*). The data shown in this figure are from the same cohort of subjects reported in the papers by Zhou et al. (2011) and Lin et al. (2012), but analyzed in different ways. Panel **a** shows the brain regions with significantly ($p < 0.001$, uncorrected; voxel size > 200) decreased gray matter density (*GMD*) in IAD subjects, as compared to normal controls. In addition to the regions reported in the original VBM paper (Zhou et al. 2011), decreased GMD was found in the *left* (−14, −9, 19; 822 voxels) and right (10, −7, 14; 962 voxels) anterior thalamus. Please note that a different statistical threshold ($p < 0.05$, with FDR correction) was used in the original paper (Zhou et al. 2011). Panel **b** shows the white matter (*WM*) tracts with abnormal microstructural integrity in IAD subjects. This figure is adapted from Fig. 2.1 of the original DTI paper (Lin et al. 2012). Panel **c** shows the results of probabilistic tractography using the segment of external capsule (*EC*) showing IAD-related FA reduction as the seed. Interestingly, the IAD-related atrophic brain regions revealed by VBM are interconnected via WM tracts showing compromised microstructural integrity. For example, the atrophic thalamus and insula are interconnected to dorsolateral prefrontal cortex (*dlPFC*) via EC and anterior limb of the internal capsule (*ALIC*)

are also frequently reported in subjects exposed to addictive substances, such as alcohol (De Bellis et al. 2008), cocaine (Lim et al. 2002, 2008; Romero et al. 2010), marijuana (Bava et al. 2009), heroin (Liu et al. 2008), ketamine (Liao et al. 2010), methamphetamine (Alicata et al. 2009; Salo et al. 2009), opiate (Bora et al. 2012; Upadhyay et al. 2010) and tobacco (Lin et al. 2013). Interestingly, we found that the pattern of WM microstructural abnormality in IAD largely resembles that which has been observed in opiate addicts (Fig. 2.4), indicating that Internet addiction may, at least in part, share similar neural mechanisms with other types of substance addiction.

Neuroimaging approaches have also been used to study brain structural changes associated with pathological gambling, a condition originally considered an impulsive-compulsive disorder, but now classified as "addictions and related disorders" in DSM-V (http://www.dsm5.org/Pages/Default.aspx). In subjects with pathological gambling, impaired WM microstructural integrity (i.e., lower FA) was found in the genu of CC, cingulum, ALIC, inferior fronto-occipital fascicle and anterior thalamic radiation (Joutsa et al. 2011). However, no volumetric differences in regional GM or WM were observed between pathological gamblers and normal controls (van Holst et al. 2012).

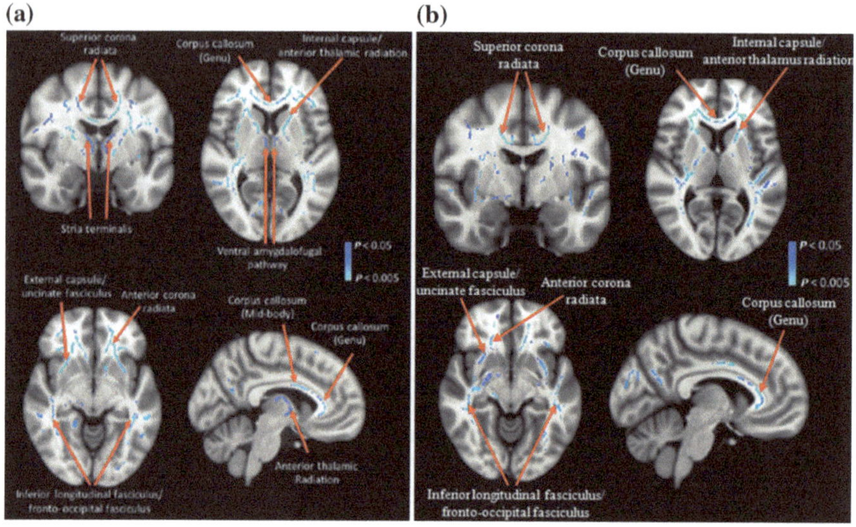

Fig. 2.4 Comparisons between the pattern of abnormal white matter integrity in Internet addiction and that in opiate addiction. The figure in panel **a** is adapted from the paper by Upadhyay et al. (2010) with permission, showing white matter tracts whose fractional anisotropy (*FA*) values were significantly ($p < 0.05$, corrected by cluster-based thresholding with $c > 3$) reduced in opiate addicts ($n = 10$), relative to normal controls ($n = 10$). The figure in panel **b** is modified from Fig. 2.1 of the paper by Lin et al. (2012). This figure is drawn in such a way that it can be directly compared with the figure shown in panel **a** by visual inspection. In this figure, the significance statistical threshold was set to $p < 0.05$ with threshold-free cluster enhancement (*TFCE*) correction. Please note that a different statistical threshold of $p < 0.01$, with TFCE correction, was used in the original paper (Lin et al. 2012). It can be seen from the figure that both the subjects with IAD and those with opiate addiction exhibited impaired microstructural integrity in the white matter tracts connecting to the prefrontal cortex, such as the genu of corpus callosum, cingulum, corona radiate, internal and external capsules. However, reduced FA in stria terminalis and ventral amygdalofugal pathway was observed only in opiate addicts, but not in subjects addicted to the Internet. These findings demonstrate that Internet addiction may, at least to some extent, share similar neural mechanisms with substance addiction

It therefore appears that IAD/OGA, substance addiction, and pathological gambling are associated with, to some extent, similar structural abnormalities in the brain, which may constitute a neural signature for the three forms of addiction.

2.4 Implications of Brain Structural Abnormalities in Internet Addiction

Overall, Internet addicts appear to have impaired microstructural integrity in the WM tracts involved in the neural circuits underlying emotion generation and processing, executive attention, decision making and cognitive control. From the research on substance addiction, we now know that the insula and PFC play crucial roles in addiction. The insula is a brain region that integrates interoceptive states into conscious feelings, and contributes mainly to the motivation or urge to use drugs and to decision-making processes that precipitate relapse (Naqvi and Bechara 2009). Among the complex functions of PFC, one of the most important ones, perhaps, is executive control, such as planning, prioritizing, organizing, and emotion processing (Chan et al. 2008). It is generally accepted that PFC abnormalities are central to the addiction-related behaviors related to executive dysfunction, including maladaptive decision-making and compulsive-repetitive behaviors (Goldstein and Volkow 2011). Subjects with IAD not only show symptoms such as craving and relapse, but also are known to have impaired abilities in impulse control (Whang et al. 2003). Through neuroimaging approaches, IAD subjects are shown to have structural abnormalities in the PFC, insula and the WM fibers connecting these two regions to other parts of brain. This may be viewed as evidence in support of the notion that impaired decision-making and executive control are important features of IAD.

Thalamus is a key target for dopamine (Sanchez-Gonzalez et al. 2005) and plays an important role in reward processing, goal-directed behaviors, as well as many other cognitive and motor functions (Corbit et al. 2003; Yu et al. 2010). Altered thalamic microstructure may contribute to the development of Internet addiction by disrupting the acquisition of stimulus-reward associations.

The human visual system is very sensitive to subtle details in movements, even to weakened stimuli such as point-light walkers (Blake and Shiffrar 2007). Excessive exposure to visual stimulation (i.e., computer monitor) may lead to neuroplastic changes in the brain regions related to visual/spatial processing, such as MOG, IOG, ITG, lingular gyrus, PCC and precuneus. Additionally, Internet addicts spend a tremendous amount of time on-line and become astonishingly skilled and accurate in mouse clicking and keyboard typing (Kuss and Griffithsemail 2012). Such training/overlearning processes may induce neuroplastic changes in the sensorimotor-related areas.

CC is the largest WM fiber tract connecting the neocortex of the two hemispheres (Delacoste et al. 1985). Reduced FA in the CC of the subjects with IAD may be indicative of alterations in the communication between the two hemispheres. The ALIC is a key region of frontal-subcortical circuits, providing connections between the thalamus/striatum and frontal cortical regions and comprising a system that plays an important role in reward and emotion processing (Mori et al. 2005). EC connects the ventral and medial prefrontal cortex to the striatum, and is involved in emotion generation and processing (Mori et al. 2005). Impaired WM integrity in the ALIC and EC may thus be indicative of alterations in reward and emotion processing in IAD. The corona radiation links the cerebral cortex to the internal capsule and provides important connections between the frontal, parietal, temporal, and occipital lobes (Wakana et al. 2004). The inferior fronto-occipital fasciculus is an association bundle connecting the frontal with the parietal and occipital lobes (Wakana et al. 2004). Impaired WM microstructure in the corona radiation and inferior fronto-occipital fasciculus is likely to be associated with the abnormal GM density/volumes in cerebral cortex.

2.5 Limitations of Previous Studies

Notwithstanding the insightful results on brain structural changes associated with IAD, a number of limitations in the previous studies need to be observed. Firstly, the diagnosis criteria used for IAD and OGA are somewhat different across studies. This may potentially result in error classification in some cases, and limit the ability to draw direct comparisons among different studies. Secondly, the structural brain imaging studies on IAD/OGA performed so far all constituted relatively small samples. Owing to this limitation, the results summarized in this chapter should to be considered preliminary, and need to be replicated in future studies with larger sample sizes. Generalization of the findings from the available studies should also be undertaken with caution. Thirdly, previous studies often used cross-sectional designs, such that the question of whether brain structural changes are a consequence or a precondition for IAD/OGA cannot be answered. Finally, IAD may have many subtypes (i.e., OGA) depending on the type of online activity one is addicted to. Furthers studies should compare the brain structural changes across different IAD subtypes to determined whether such changes are specific to on-line activities, or caused by uncontrolled use of the Internet per se.

2.6 Summary and Perspectives

Taken together, the structural MRI studies available thus far clearly demonstrated that Internet addiction is associated with anatomical abnormalities in the brain involving both GM and WM. Reduced GM density/volume and cortical thickness are consistently observed in the PFC and insula of the subjects addicted to the Internet, who also showed impaired microstructural integrity in the WM tracts that connect to the PFC. The brain regions showing structural changes in IAD are known to be involved in reward, emotion generation and processing, executive attention, decision making and cognitive control. The pattern of IAD-related structural abnormalities in the brain is also shown to be similar, to some extent, to that observed in substance addiction. It therefore may be concluded that Internet addiction may, at least in part, share similar neural mechanisms with substance addiction and pathological gambling. However, only eight structural brain imaging studies on IAD have been published to date. Further studies, especially longitudinal studies with large sample sizes, are needed to elucidate the exact relationship between uncontrolled use of the Internet and plastic structural changes in the brain.

Acknowledgments This work was partially supported by grants from National Basic Research Program of China (973 Program, 2011CB707802) and Natural Science Foundation of China (81171302 and 21221064).

References

Aboujaoude E (2010) Problematic internet use: an overview. World Psychiatry 9:85–90
Alicata D, Chang L, Cloak C et al (2009) Higher diffusion in striatum and lower fractional anisotropy in white matter of methamphetamine users. Psychiatry Res Neuroimaging 174:1–8
Ashburner J, Friston KJ (2000) Voxel-based morphometry—the methods. Neuroimage 11:805–821
Barros-Loscertales A, Garavan H, Bustamante JC et al (2011) Reduced striatal volume in cocaine-dependent patients. Neuroimage 56:1021–1026
Basser PJ, Mattiello J, Lebihan D (1994a) Estimation of the effective self-diffusion tensor from the NMR spin-echo. J Magn Reson 103:247–254
Basser PJ, Mattiello J, Lebihan D (1994b) MR diffusion tensor spectroscopy and imaging. Biophys J 66:259–267
Basser PJ, Pajevic S, Pierpaoli C et al (2000) In vivo fiber tractography using DT-MRI data. Magn Reson Med 44:625–632
Basser PJ, Pierpaoli C (1996) Microstructural and physiological features of tissues elucidated by quantitative-diffusion-tensor MRI. J Magn Reson 111:209–219
Bava S, Frank LR, McQueeny T et al (2009) Altered white matter microstructure in adolescent substance users. Psychiatry Res Neuroimaging 173:228–237
Beard KW, Wolf EM (2001) Modification in the proposed diagnostic criteria for internet addiction. CyberPsychol Behav 4:377–383
Bernardi S, Pallanti S (2009) Internet addiction: a descriptive clinical study focusing on comorbidities and dissociative symptoms. Compr Psychiatry 50:510–516
Blake R, Shiffrar M (2007) Perception of human motion. Annu Rev Psychol 58:47–73
Bora E, Yucel M, Fornito A et al (2012) White matter microstructure in opiate addiction. Addict Biol 17:141–148
Chakraborty K, Basu D, Vijaya Kumar KG (2010) Internet addiction: consensus, controversies, and the way ahead. East Asian Arch Psychiatry 20:123–132
Chan RCK, Shum D, Toulopoulou T et al (2008) Assessment of executive functions: review of instruments and identification of critical issues. Arch Clin Neuropsychol 23:201–216
Conturo TE, Lori NF, Cull TS et al (1999) Tracking neuronal fiber pathways in the living human brain. Proc Natl Acad Sci 96:10422–10427
Corbit LH, Muir JL, Balleine BW (2003) Lesions of mediodorsal thalamus and anterior thalamic nuclei produce dissociable effects on instrumental conditioning in rats. Eur J Neurosci 18:1286–1294
De Bellis MD, Van Voorhees E, Hooper SR et al (2008) Diffusion tensor measures of the corpus callosum in adolescents with adolescent onset alcohol use disorders. Alcohol Clin Exp Res 32:395–404
Delacoste MC, Kirkpatrick JB, Ross ED (1985) Topography of the human corpus callosum. J Neuropathol Exp Neurol 44:578–591
Dell'Acqua F, Catani M (2012) Structural human brain networks: hot topics in diffusion tractography. Curr Opin Neurol 375–383
Dong G, DeVito E, Huang J et al (2012) Diffusion tensor imaging reveals thalamus and posterior cingulate cortex abnormalities in internet gaming addicts. J Psychiatr Res 46:1212–1216
Douglas AC, Mills JE, Niang M et al (2008) Internet addiction: meta-synthesis of qualitative research for the decade 1996–2006. Comput Hum Behav 24:3027–3044
Ersche KD, Williams GB, Robbins TW et al (2013) Meta-analysis of structural brain abnormalities associated with stimulant drug dependence and neuroimaging of addiction vulnerability and resilience. Curr Opin Neurobiol 23:615–624
Franklin TR, Acton PD, Maldjian JA et al (2002) Decreased gray matter concentration in the insular, orbitofrontal, cingulate, and temporal cortices of cocaine patients. Biol Psychiatry 51:134–142

Goldstein RZ, Volkow ND (2011) Dysfunction of the prefrontal cortex in addiction: neuroimaging findings and clinical implications. Nat Rev Neurosci 12:652–669
Han DH, Lyoo IK, Renshaw PF (2012) Differential regional gray matter volumes in patients with online game addiction and professional gamers. J Psychiatr Res 46:507–515
Holden C (2001) 'Behavioral' addictions: do they exist? Science 294:980–982
Hong SB, Kim JW, Choi EJ et al (2013) Reduced orbitofrontal cortical thickness in male adolescents with internet addiction. Behav Brain Funct 9:11
Jones DK, Symms MR, Cercignani M et al (2005) The effect of filter size on VBM analyses of DT-MRI data. Neuroimage 26:546–554
Joutsa J, Saunavaara J, Parkkola R et al (2011) Extensive abnormality of brain white matter integrity in pathological gambling. Psychiatry Res Neuroimaging 194:340–346
Kim SJ, Lyoo IK, Hwang J et al (2006) Prefrontal grey-matter changes in short-term and long-term abstinent methamphetamine abusers. Int J Neuropsychopharmacol 9:221–228
Ko CH, Yen JY, Yen CF et al (2012) The association between internet addiction and psychiatric disorder: a review of the literature. Eur Psychiatry 27:1–8
Kuss DJ, Griffiths MD, Karila L et al (2013) Internet addiction: a systematic review of epidemiological research for the last decade. Curr Pharm Des
Kuss DJ, Griffithsemail MD (2012) Internet and gaming addiction: a systematic literature review of neuroimaging studies. Brain Sci 2:347–374
Le Bihan D (2003) Looking into the functional architecture of the brain with diffusion MRI. Nat Rev Neurosci 4:469–480
Lerch JP, Pruessner JC, Zijdenbos A et al (2005) Focal decline of cortical thickness in Alzheimer's disease identified by computational neuroanatomy. Cereb Cortex 15:995–1001
Liao YH, Tang JS, Liu TQ et al (2012) Differences between smokers and non-smokers in regional gray matter volumes: a voxel-based morphometry study. Addict Biol 17:977–980
Liao YH, Tang JS, Ma MD et al (2010) Frontal white matter abnormalities following chronic ketamine use: a diffusion tensor imaging study. Brain 133:2115–2122
Lim KO, Choi SJ, Pomara N et al (2002) Reduced frontal white matter integrity in cocaine dependence: a controlled diffusion tensor imaging study. Biol Psychiatry 51:890–895
Lim KO, Wozniak JR, Mueller BA et al (2008) Brain macrostructural and microstructural abnormalities in cocaine dependence. Drug Alcohol Depend 92:164–172
Lin FC, Wu GY, Zhu L et al (2013) Heavy smokers show abnormal microstructural integrity in the anterior corpus callosum: a diffusion tensor imaging study with tract-based spatial statistics. Drug Alcohol Depend 129:82–87
Lin FC, Yu CS, Jiang TZ et al (2006) Quantitative analysis along the pyramidal tract by length-normalized parameterization based on diffusion tensor tractography: application to patients with relapsing neuromyelitis optica. Neuroimage 33:154–160
Lin FC, Zhou Y, Du YS et al (2012) Abnormal white matter integrity in adolescents with internet addiction disorder: a tract-based spatial statistics study. PLoS ONE 7:e30253
Liu H, Li L, Hao Y et al (2008) Disrupted white matter integrity in heroin dependence: a controlled study utilizing diffusion tensor imaging. Am J Drug Alcohol Abuse 34:562–575
Liu HH, Hao YH, Kaneko Y et al (2009) Frontal and cingulate gray matter volume reduction in heroin dependence: optimized voxel-based morphometry. Psychiatry Clin Neurosci 63:563–568
Lyoo IK, Pollack MH, Silveri MM et al (2006) Prefrontal and temporal gray matter density decreases in opiate dependence. Psychopharmacology 184:139–144
MacDonald D, Kabani N, Avis D et al (2000) Automated 3-D extraction of inner and outer surfaces of cerebral cortex from MRI. Neuroimage 12:340–356
Makris N, Gasic GP, Kennedy DN et al (2008a) Cortical thickness abnormalities in cocaine addiction—a reflection of both drug use and a pre-existing disposition to drug abuse? Neuron 60:174–188
Makris N, Oscar-Berman M, Jaffin SK et al (2008b) Decreased volume of the brain reward system in alcoholism. Biol Psychiatry 64:192–202

McIntosh AM, Maniega SM, Lymer GKS et al (2008) White matter tractography in bipolar disorder and schizophrenia. Biol Psychiatry 64:1088–1092
Mechelli A, Price CJ, Friston KJ et al (2005) Voxel-based morphometry of the human brain: methods and applications. Curr Med Imaging Rev 1:105–113
Morahan-Martin J (2005) Internet abuse—addiction? Disorder? Symptom? Alternative explanations? Soc Sci Comput Rev 23:39–48
Mori S, Crain BJ, Chacko VP et al (1999) Three-dimensional tracking of axonal projections in the brain by magnetic resonance imaging. Ann Neurol 45:265–269
Mori S, Wakana S, Nagae-Poetscher LM et al (2005) MRI atlas of human white matter. Elsevier, San Diego
Naqvi NH, Bechara A (2009) The hidden island of addiction: the insula. Trends Neurosci 32:56–67
Pierpaoli C, Jezzard P, Basser PJ et al (1996) Diffusion tensor MR imaging of the human brain. Radiology 201:637–648
Romero MJ, Asensio S, Palau C et al (2010) Cocaine addiction: diffusion tensor imaging study of the inferior frontal and anterior cingulate white matter. Psychiatry Res Neuroimaging 181:57–63
Salat DH, Buckner RL, Snyder AZ et al (2004) Thinning of the cerebral cortex in aging. Cereb Cortex 14:721–730
Salo R, Nordahl TE, Buonocore MH et al (2009) Cognitive control and white matter callosal microstructure in methamphetamine-dependent subjects: a diffusion tensor imaging study. Biol Psychiatry 65:122–128
Sanchez-Gonzalez MA, Garcia-Cabezas MA, Rico B et al (2005) The primate thalamus is a key target for brain dopamine. J Neurosci 25:6076–6083
Schwartz DL, Mitchell AD, Lahna DL et al (2010) Global and local morphometric differences in recently abstinent methamphetamine-dependent individuals. Neuroimage 50:1392–1401
Smith SM, Jenkinson M, Johansen-Berg H et al (2006) Tract-based spatial statistics: voxelwise analysis of multi-subject diffusion data. Neuroimage 31:1487–1505
Sowell ER, Peterson BS, Kan E et al (2007) Sex differences in cortical thickness mapped in 176 healthy individuals between 7 and 87 years of age. Cereb Cortex 17:1550–1560
Upadhyay J, Maleki N, Potter J et al (2010) Alterations in brain structure and functional connectivity in prescription opioid-dependent patients. Brain 133:2098–2114
van Holst RJ, de Ruiter MB, van den Brink W et al (2012) A voxel-based morphometry study comparing problem gamblers, alcohol abusers, and healthy controls. Drug Alcohol Depend 124:142–148
Wakana S, Jiang HY, Nagae-Poetscher LM et al (2004) Fiber tract-based atlas of human white matter anatomy. Radiology 230:77–87
Weng CB, Qian RB, Fu XM et al (2013) Gray matter and white matter abnormalities in online game addiction. Eur J Radiol 82:1308–1312
Whang LS, Lee S, Chang G (2003) Internet over-users' psychological profiles: a behavior sampling analysis on internet addiction. CyberPsychol Behav 6:143–150
Yen JY, Ko CH, Yen CF et al (2007) The comorbid psychiatric symptoms of Internet addiction: attention deficit and hyperactivity disorder (ADHD), depression, social phobia, and hostility. J Adolesc Health 41:93–98
Young KS (1996) Psychology of computer use: XL. Addictive use of the Internet: a case that breaks the stereotype. Psychol Rep 79:899–902
Young KS (1998) Internet addiction: the emergence of a new clinical disorder. CyberPsychol Behav 1:237–244
Yu C, Gupta J, Yin HH (2010) The role of mediodorsal thalamus in temporal differentiation of reward-guided actions. Front Integr Neurosci 4:14
Yuan K, Cheng P, Dong T et al (2013) Cortical thickness abnormalities in late adolescence with online gaming addiction. PLoS ONE 8:e53055

Yuan K, Qin W, Dong MH et al (2010) Gray matter deficits and resting-state abnormalities in abstinent heroin-dependent individuals. Neurosci Lett 482:101–105
Yuan K, Qin W, Wang G et al (2011) Microstructure abnormalities in adolescents with internet addiction disorder. PLoS ONE 6:e20708
Zhang XC, Salmeron BJ, Ross TJ et al (2011) Factors underlying prefrontal and insula structural alterations in smokers. Neuroimage 54:42–48
Zhou Y, Lin FC, Du YS et al (2011) Gray matter abnormalities in Internet addiction: a voxel-based morphometry study. Eur J Radiol 79:92–95

Chapter 3
Functional Imaging of Internet Gaming Disorder

Chih-Hung Ko, Gin-Chung Liu and Ju-Yu Yen

Abstract The present chapter reviews functional magnetic resonance imaging (fMRI) studies of Internet gaming disorder (IGD) and Internet addiction (IA). The review includes reports of the use of fMRI, including arterial spin-labeled fMRI for analyzing task-related performance, resting functional connectivity, and effective connectivity. However, any conclusions regarding the mechanisms of IGD or IA would be premature due to the many limitations of previous studies. Future work should consider the diagnostic criteria outlined in the DSM-5 when recruiting subjects with IGD. Further fMRI studies with adequate sample sizes of at least 20 participants, reasonable hypotheses, effective designs, and precise data analyses by integrated research teams are needed for further elucidation of the mechanisms of IGD and IA.

C.-H. Ko (✉) · J.-Y. Yen
Department of Psychiatry, Kaohsiung Medical University Hospital, Kaohsiung Medical University, Kaohsiung, Taiwan
e-mail: chihhungko@gmail.com

C.-H. Ko · J.-Y. Yen
Department of Psychiatry, Faculty of Medicine, College of Medicine, Kaohsiung Medical University, Kaohsiung, Taiwan

C.-H. Ko
Department of Psychiatry, Kaohsiung Municipal Hsiao-Kang Hospital, Kaohsiung Medical University, Kaohsiung, Taiwan

G.-C. Liu
Department of Medical Imaging, Kaohsiung Medical University Hospital, Kaohsiung Medical University, Kaohsiung, Taiwan

G.-C. Liu
Department of Medical Imaging, Faculty of Medicine, College of Medicine, Kaohsiung Medical University, Kaohsiung, Taiwan

J.-Y. Yen
Department of Psychiatry, Kaohsiung Municipal Ta-Tung Hospital, Kaohsiung, Taiwan

C. Montag and M. Reuter (eds.), *Internet Addiction*, Studies in Neuroscience, Psychology and Behavioral Economics, DOI 10.1007/978-3-319-07242-5_3

3.1 Introduction

Internet gaming is now among the most popular recreational activities worldwide. While the Internet enables access to online games 24 h a day, smart phones further enable access from any location. However, loss of control over the time spent engaged in online gaming may have negative consequences. Owing to the significant difference between generalized internet addiction and internet gaming disorder (Davis 2001; Montag et al. 2014) and the potentially adverse impacts of IGD on the health of internet users, the Diagnostic and Statistical Manual of Mental Disorders Fifth Edition (DSM-5) proposed diagnostic criteria specific for internet gaming disorder (IGD) in 2013. It defined addiction to Internet gaming among the conditions for further study of Sect. 3.3, and suggested that more evidence was needed before IGD can be included as a standard disorder in the DSM system (American Psychiatric Association; APA 2013). Several neurobiological and neurocognitive studies have evaluated the neuropsychological mechanisms of IGD (e.g. Dong et al. 2012b; Han et al. 2010; Ko et al. 2009a). Although some of the proposed clinical presentations and diagnostic criteria for IGD resemble those for substance addiction (Ko et al. 2006, 2012), no studies have conclusively determined whether IGD shares similar mechanisms with substance use disorder. Since neurobiological mechanisms of substance use disorder have been extensively researched in the past decade (Volkow et al. 2010), studying the neurobiological mechanisms of IGD might reveal whether they resemble those in substance use disorder (Table 3.1).

The present chapter reviews functional magnetic resonance imaging (fMRI) studies of internet gaming disorder and internet addiction (IA). The review includes any reports of the use of fMRI, including arterial spin labeled fMRI, for analyzing task-related performance, resting functional connectivity, or effective connectivity.

3.2 Functional Magnetic Resonance Imaging

fMRI is a functional technique used to record cerebral hemodynamic changes during a specific task to demonstrate possible brain mechanisms of a specific neurocognitive function (Tejado Lde et al. 2010). Analyses of blood-oxygen level dependent (BOLD) contrast, i.e., the difference in magnetic susceptibility between oxyhemoglobin and deoxyhemoglobin, enable indirect study of hemodynamic changes based on variation in magnetic signals (Huettel et al. 2009). By comparing MR signals obtained while the subject is resting and while the subject is performing a specific task, this technique can reveal hemodynamic changes related to a specific neuro-cognitive response.

For evaluating mechanisms of psychiatric disorders, fMRI is preferable to PET and SPECT because it does not require radiation exposure. Other advantages include its wider availability and lower expense. Moreover, fMRI provides superior spatial and temporal resolution (Tejado Lde et al. 2010). However, isolating the hemodynamic change associated with a specific stimulus is very difficult. The

Table 3.1 Summary of fMRI studies of internet gaming disorder (IGD) or internet addiction (IA)

Author (year)	Definition	Subject	Task or aim of study	Reaction difference	Multiple correction
Ko et al. (2009a)	DCIA by psychiatric interview; addicted to the same game	10 male adults with IGD and 10 controls	Block design, gaming cue induced reactivity	Right or bito-frontal, DLPFC, nucleus accumbens, caudate, and bilateral anterior cingulate	$P < 0.0005$ and >50 voxels
Ko et al. (2013)	DCIA by psychiatric interview; addicted to the same game	15 male adults with IGD, 15 with remitted IGD, and 15 controls	Event related, gaming cue induced reactivity	Bilateral DLPFC, precuneus, left posterior cingulate, parahippocampus and right anterior cingulate	$P < 0.001$ and cluster size >10 voxels
				Right DLPFC and left parahippocampus in comparison to remission group	
Han et al. (2010)	IAT (50 or higher) with clinical functional impairment; addicted to the same game	11 male adults with IGD under bupropion treatment and 8 controls	Block design, gaming cue induced reactivity	Left occipital lobe, DLPFC, and parahippocampus	$p < 0.05$ in false discovery rate correction
Lorenz et al. (2013)	3 of 6 criteria for computer game addiction; addicted to the same game	8 male adults with IGD and 9 controls	Short presentation and long presentation dotprobe task; psychological interaction analysis for functional connectivity	Right hippocampus and right inferior frontal gyrus	AlphaSim correction
Ko et al. (2013b)	DCIA by psychiatric interview; addicted to the same game	16 male adults with concurrent IGD and nicotine dependence; 16 controls	Event related, gaming cue induced reactivity; conjugation analysis	Bilateral parahippocampal gyrus	$p < 0.05$ with small volume correction

(continued)

Table 3.1 (continued)

Dong et al. (2012b)	IAT (cutoff: 80), and Chinese IAT	12 male adults with IGD and 12 controls	Event related, stroop task	Left posterior cingulate, bilateral anterior cingulate	$p < 0.05$ in false discovery rate correction
	IAT (cut/off: 80), and Chinese IAT	12 male adults with IGD and 12 controls	Reality-simulated guessing task	High orbitofrontal cortex in win condition, and low anterior cingulate in loss condition	$p < 0.05$ in false discovery rate correction
Kim et al. (2012)	KAIAS (50 or higher)	12 male adults with IGD and 12 controls	Ball throwing animation task for disembodiment state	High brain activation of disembodiment near the left temporo-parieto-occipital junction and right parahippocampal area	$p < 0.0005$ and $k > 10$
Dong et al. (2012a)	IAT (cut/off: 80)	15 male adults with IGD and 14 controls	Resting fMRI; regional homogeneity combined with a seed region functional connectivity	Enhanced ReHo in brain stem, inferior parietal lobe, left posterior cerebellum, and left middle frontal gyrus; no significant difference in seed region analysis of inferior parietal lobe	$p < 0.05$ FDR correction
Liu et al. (2010)	Modified Young diagnostic questionnaire	19 adults (11 males, 8 females) with IGD and 19 controls	Resting fMRI; regional homogeneity	Cerebellum, brainstem, right cingulate gyrus, bilateral parahippocampus, right frontal lobe, left superior frontal gyrus, left precuneus, right inferior temporal gyrus, left superior temporal gyrus and middle temporal gyrus	$p < 0.001$ and a minimum cluster size of 270 mm^3

(continued)

Table 3.1 (continued)

Hong et al. (2013)	IAT (cut/off: 50)	12 adolescents with IA and 11 controls	Resting fMRI; inter-regional connectivity	Most impaired connections involved subcortical brain region	False discovery rate correction
Ding et al. (2013)	Modified Young diagnostic questionnaire	17 adolescents (13 males, 4 females) with IGD and 24 controls (16 male, 8 female)	Resting fMRI; default network as seed region of posterior cingulate	Increased FC in the bilateral cerebellum, posterior lobe, and middle temporal gyrus; decreased bilateral inferior parietal lobe and right inferior temporal gyrus	AlphaSim correction
	Modified Young diagnostic questionnaire	15 adolescents (13 males, 2 females) with IGD and 18 controls (14 male, 4 female)	Arterial spin-labeled perfusion magnetic resonance imaging for baseline cerebral blood perfusion	Increased CBF in the left inferior temporal lobe, left parahippocampal gyrus/amygdala, right medial frontal lobe/anterior cingulate, bilateral insula, right middle temporal gyrus, right precentral gyrus, left supplementary motor area, left cingulate, and right inferior parietal lobe	AlphaSim correction

DCIA diagnostic criteria for internet addiction proposed by Ko et al. (2009a, b); *IGD* Internet gaming disorder; *DLPFC* dorsolateral prefrontal cortex; *IAT* Internet addiction test developed by Yang (1998); *KAIAS* Korean adolescent internet addiction scale (Kim 2003); *IA* Internet addiction not specific to online gaming

paradigm must be well designed to avoid artifacts and misinterpretation. Further, since the BOLD signal provides only an indirect estimation of cerebral blood flow (CBF), the fMRI should be interpreted cautiously. The combination of fMRI techniques with insights from PET provides more direct insights to the biochemical mechanism of human behaviors.

3.3 Use of fMRI for Studying IGD

fMRI is among the most important imaging tools for investigating mechanisms of addiction, including response to a substance, vulnerability to addiction, characteristics or symptoms of addictive behavior, and consequences of addiction (Fowler et al. 2007). As fMRI is widely available in industrialized countries, which tend to have the highest prevalence of IGD and IA, the present chapter reviews fMRI results reported in brain studies of subjects diagnosed with IGD. The fMRI results should be interpreted cautiously since the hypotheses in some of these studies have not been supported by further psychopathology, cognition, or physiology studies. Additionally, before beginning a new fMRI study, a rational hypothesis must be derived based on previous evidence obtained in brain imaging studies. Therefore, a literature review of applications of fMRI for studying IGD or IA is timely.

3.3.1 Task-Related fMRI Study

A task-related fMRI study typically uses visual, auditory, or other stimuli, to provoke two or more different cognitive processes in the subject. The typical design has two conditions, an experimental condition and a control condition. The goal of evaluation is to test the hypothesis that the signal differs between the two conditions. The trials are designed to alternate between the experimental and control condition (Glover 2011).

Classic experimental designs in fMRI represent the block or event related design. In block design, each block will have a duration of a certain number of fMRI scans, about 20–30 s, and within each block only one condition (such as pictures from a computer game) is presented. Although the block design is considered optimal for detecting activation, numerous cognitive processes may occur within 20–30 s. Therefore, other approaches, such as the event-related-design, are also used in fMRI studies. Event-related designs enable the researcher to detect changes in the BOLD hemodynamic response in response to specific events. Event-related designs are presented in a randomized sequence and the between-stimuli duration is varied. Therefore, an event-related design is better able to characterize the timing of the change in amplitude of the hemodynamic response in form of the BOLD signal.

For example, in the Go/Nogo task, an fMRI study with an event-related design can distinguish the BOLD response to the Go, Nogo, and failed inhibited trials.

Therefore, an fMRI study is preferable for assessing the functional anatomy of response inhibition and error processing (Criaud and Boulinguez 2013). Both the block design and event-related design have been used to study IGD, particularly in the cue-induced reactivity paradigm. Such paradigms confront IGD patients with pictures from their favorite computer game while they are in an fMRI setting and their brain activity is being recorded. These designs have also been used to study reward sensitivity, inhibitory control, and disembodiment.

3.3.1.1 The fMRI for Analyzing Cue-Induced Reactivity in IGD or IA (Ko et al. 2009a)

Early studies of cue-induced gaming urge with block designs
Ko et al. (2009a) recruited 10 adults who met the diagnostic criteria for IGD according to psychiatric interviews, and 10 healthy controls without IGD. All subjects in the experimental group were addicted to the same online game. This block-design study of the cue-induced craving paradigm performed fMRI scans while the subjects were shown screenshots of video games. The significance threshold was set to $p < 0.0005$, and the cluster size was >50 voxels.
The experimental results revealed that, compared to controls, the IGD group had higher activations in the right orbito-frontal cortex, bilateral anterior cingulate, right dorsolateral prefrontal cortex (DLPFC), right nucleus accumbens, and right caudate nucleus. The authors suggested that these areas may be considered neural substrates of the cue-induced gaming urge in IGD. The results also indicated that cue-induced brain reactivity in IGD resembles that in substance use disorder.

The limitations of the study were the limited number of subjects and the use of a block design. Further, the *p* value was not consistently corrected for multiple comparisons.

Brain correlates of cue-induced gaming urge and the course of IGD (Ko et al. 2013a)
Ko et al. (2013a) recruited 15 adults with IGD, 15 remitted subjects, and 15 healthy controls. Internet gaming disorder was diagnosed according to the modified diagnostic criteria of IA (DCIA) observed in psychiatric interviews (Ko et al. 2009b). All subjects recruited for the study had been diagnosed with addiction to the same online game for at least 1 year. All subjects in the remission group had a history of addiction to the same online game, and all had been in remission from IGA for at least 6 months, according to DCIA criteria. In this event-related study of the cue-induced craving paradigm, subjects viewed screenshots of online gaming activity while undergoing fMRI scans. The significance threshold in the between-group analyses was $p < 0.001$, and the cluster size was >10 voxels.

Compared to the control group, the IGD group showed higher activations in the bilateral DLPFC, precuneus, left posterior cingulate, parahippocampus and right anterior cingulate in response to gaming cues. The authors hypothesized that these brain areas are associated with the gaming craving activated by cue exposure.

The authors suggested that the activation pattern was consistent with the model of substance use disorder developed earlier by Volkow et al. (2010). Furthermore, activation of the left superior parietal lobe in response to gaming cues was higher in the remission group than in the control group. Lastly, compared with the remission group, the IGA group had higher activation over the right DLPFC (BA46), left parahippocampus (BA19) and left middle temporal gyrus (BA 39). Therefore, the authors suggested that the DLPFC and parahippocampus are potential markers of cue-induced brain activation in subjects currently in a state of addiction to online gaming.

This study was also limited by the relatively low sample size, and the lack of standard corrections of the p value for multiple comparisons. In the context of the aforementioned studies, the findings by Montag et al. (2012) are of interest. First, they reported higher activation of the lateral frontal cortex in control subjects compared to gamers, when processing screen-shots of unpleasant pictures from the International Affective Picture System. The authors suggested that the higher brain activity in the (left) lateral prefrontal cortex could represent a protection mechanism against experiencing negative emotions by down-regulating limbic brain activity. In addition, this study compared the processing of stills from the first-person-shooter video game Counterstrike (contrast Counterstrike > neutral pictures), in both gamers and controls. The group of gamers was associated with stronger DLPFC activation while processing these pictures, which might reflect script activation on how to react in the computer game. Of note, no elevated activation in striatal regions was observed in this group of excessive gamers while processing the pictures from the favorite game.

Gaming cue-induced reactivity in the course of Bupropion treatment (Han et al. 2010)

Han et al. (2010) recruited 11 adults diagnosed with IGD and eight healthy controls. The IGD diagnoses were based on a self-report questionnaire combined with a score of at least 50 on the IA test (Widyanto and McMurran 2004; Young 1998). In this block design study, all subjects viewed online gaming cues while undergoing fMRI scans. The significance threshold for a correction in the False Discovery Rate for 100 adjacent voxels was set to $p < 0.05$.

The results showed that, compared to controls, subjects with IGD had higher brain activation in response to gaming cues over the left occipital lobe, DLPFC, and parahippocampus. However, after 6 weeks of treatment with bupropion extended-release tablets (SR), the subjects showed significant decreases in β values over the DLPFC ($p = 0.04$) and in craving scores ($p = 0.04$). Therefore, the authors suggested that the effects of bupropion SR on brain activity of craving for online gaming among subject with IGD.

Cue-induced reactivity in dot probe task and connectivity analysis (Lorenz et al. 2013)

Lorenz et al. (2013) recruited eight adult males with IGD and nine healthy controls. The recruitment criteria for the experimental group of males with IGD were at least three of the following six criteria for IGA within the last 12 months: craving, impaired control of playing, withdrawal, development of tolerance, progressive neglect of other pleasure, and playing despite harmful consequences. The healthy controls were casual computer game players who were familiar with WoW, World of Warcraft, the online role player game. All subjects underwent fMRI scans while viewing screenshots of WoW in short presentation and long presentation trials of the dot probe task. The dot probe task is designed to assess attention bias. Participants are asked to stare at a fixation cross on the center of the screen. Two stimuli, one of which is neutral and one of which is WoW related, appear randomly on either side of the screen. The stimuli are presented for a short or long duration, before a dot is presented in the location of one former stimulus. Participants are instructed to indicate the location of this dot as quickly as possible.

In the dot probe paradigm, subjects with IGD generally show attentional bias toward stimuli with a positive valence. Analyses of the fMRI data revealed that, compared to the healthy control group, the experimental group had a different activation in short presentation trials {with the contrast computed as [(WoW > neutral) > (positive emotion picture > neutral)]} in areas known to be associated with craving in addiction (e.g., ventral visual path, right hippocampus, and right inferior frontal gyrus). However, this activation pattern was only observed when stimuli were presented for durations shorter than 2 s. The stimulus duration related modulation of connectivity strength from the right inferior gyrus to cue reactivity related regions suggested the presence of an inhibiting effect during the long presentation trials.

The authors further hypothesized that subjects with IGD might reveal inhibition of motivation-related brain regions in long presentation trials but not in short presentation trials. An explorative post hoc psychological interaction analysis was used to test for altered functional connectivity between the right inferior frontal gyrus and other brain areas. The analysis revealed significant group differences in the presentation duration dependent modulation of coupling strength between the right inferior gyrus and areas related to cue reactivity, which was consistent with the hypothesis of the authors of that study.

However, these results should be interpreted cautiously because of the small sample size. Furthermore, the complex contrast [(WoW > neutral) > (positive emotion picture > neutral)] made it difficult to explain the significant activation. That is, the significant difference in subtraction might possibly result from, not only the higher response to the WoW pictures, but also the lower response to the positive emotion pictures.

Comparison of cue-induced brain reactivity between IGD and nicotine use disorder in the same subjects

Ko et al. (2013b) recruited an experimental group of 16 adults with concurrent IGD and nicotine dependence and 16 healthy controls. The recruitment criteria for the experimental group were psychiatric interview results indicating that the subject met both the diagnostic criteria for Internet addiction (Ko et al. 2009b) and the DSM-IV criteria for nicotine dependence (APA 2000). Additionally, all subjects in the experimental group were currently addicted to the same game. In contrast, the healthy controls had no history of IGD or nicotine dependence. All subjects then underwent fMRI scans while viewing screenshots of computer games, photographs associated with cigarette smoking, and neutral stimuli. The selection of stimuli in this event-related design was based on the cue-induced reactivity paradigm. Full factorial analysis was used to compare comorbid subjects and controls in terms of brain activation induced by gaming or smoking cues. The primary factor represented the comorbid group versus the control group (group effect) while the secondary repeated measures factor represented the gaming cue reactivity versus the smoking cue reactivity (cue effect). The group effect for gaming or smoking cue reactivity was considered significant at $p < 0.05$ with a small volume correction in the region of interest. The analysis identified the brain correlates of gaming urge (game cue reactivity of the comorbid group—that of the control group) and smoking craving (smoking cue reactivity of the comorbid group—that of the control group). Conjunction analysis is defined as "the joint refutation of multiple null hypotheses" (Friston et al. 2005). That is, it identifies a significant difference in the activation of a brain area during two different tasks. Conjunction analysis with conjunction null hypothesis (Nichols et al. 2005) was performed ($p < 0.05$ with small volume correction) to identify the brain correlates common to both gaming and smoking.

The results of this event-related design showed that, in response to gaming cues, activations over the bilateral parahippocampal gyrus, precuneus, left DLPFC, and anterior cingulate were significantly higher in the IGD group than in controls. The literature suggests that the brain correlates of cue-induced gaming urge resemble those of cue-induced substance craving (Han et al. 2010; Ko et al. 2009, 2013a, b). However, these conclusions are not based on a direct comparison between BOLD response to gaming urge and substance craving in addicts craving for the Internet or nicotine only. Furthermore, even if a study had used a similar design to compare the BOLD response to gaming use of IGD subjects and that to substance craving of drug abuser, the comparison might have been biased by differences in subject characteristics between two disorders. The present study included a comorbid group to enable comparisons of brain activation between cue-induced gaming urge and cue-induced smoking craving in the same brain. The comparisons showed that, before subtracting the reaction of the control group, the gaming cue activated a brain pattern similar to that activated by the smoking cue in the comorbid group. After subtracting the reaction of the control group, the comorbid group showed that cue-induced gaming urge and cue-induced smoking craving activated the parahippocampus and the anterior cingulate. A further conjunction analysis showed significant activation of the bilateral parahippocampal gyrus by both the gaming urge and smoking craving. Therefore, the parahippocampus may be associated with mechanisms of cue-induced brain activities common to both IGA and nicotine dependence.

Similar to the previous research, this study is also limited by the small number of subjects. Including a group with only IGD and a group with only nicotine dependence might have provided stronger support for the conclusions of the study. Lastly, the characteristics of the comorbid group might have differed from those of subjects who only have IGD or subjects who only have nicotine dependence. Specific characteristics of the group might have biased the interpretation of the results.

Summary of gaming cue-induced reactivity

Despite the different designs in the aforementioned studies of gaming cue-induced reactivity, they all reported cue-induced reactivity over the parahippocampus, anterior cingulate, precuneus, and DLPFC (Han et al. 2010; Ko et al. 2009, 2013a, b; Lorenz et al. 2013). These consistent results indicate that these areas participate in brain reactivity to the cue-induced gaming urge.

The literature agrees that the parahippocampus has a role in cue-induced craving in substance use disorder (Skinner and Aubin 2010). The parahippocampus receives input from the nucleus accumbens and amygdala and evaluates the behavioral significance of sensory information (Salzmann et al. 1993). It also provides a contextual representation function and is an important afferent pathway to the hippocampus (Rudy 2009). Thus, it may contribute to the emotional response produced by cues such as screenshots of online gaming activity. Exposure to gaming cues causes the hippocampus to produce an emotional response based on contextual memory. By integrating contextual representations with their emotional significance, the parahippocampus contributes to the craving for an online gaming experience. This may explain why both the gaming urge and smoking craving cause a strong activation of the bilateral parahippocampal gyrus (Ko et al. 2013b).

The DLPFC contributes to the evaluation and integration of emotion-related information received via sensory input and affective information received from the amygdale and nucleus accumbens (Skinner and Aubin 2010). Another role of the DLPFC is generating emotionally rewarding responses to decision-making (Mitchell 2011). Thus, the DLPFC contributes to goal-directed actions by integrating information from other brain regions (Goldman-Rakic 2002). In substance addiction, the presence of substance cues and the generation of positive expectancy cause the DLPFC to maintain the craving response (Bonson et al. 2002). Cue-induced activation of the DLPFC is reportedly decreased by bupropion treatment (Han et al. 2010) and is increased in subjects of IGD (Ko et al. 2013a, b). Together, these findings indicate that gaming cue-induced activation of the DLPFC is a potential marker of the addiction state. Because of its role in the decision to act out addictive behavior, the DLPFC is a candidate marker of the current addiction state.

The precuneus is associated with visual imagery, attention, and memory retrieval and participates in visual processes by integrating images with related memories (Cavanna and Trimble 2006). The anterior cingulate area of the brain is believed to contribute to the functional brain activation associated with craving for addictive substances (Heinz et al. 2009; Kilts et al. 2001). It participates in attention and memory processes by encoding the motivational value of stimuli (Heinz et al. 2009). The anterior cingulate is also involved in the salience of emotions, motivational information, and regulatory control over reward-seeking behavior (Chiamulera 2005; Risinger et al. 2005). That is, when an affective response is activated by gaming cues, the anterior cingulate contributes to whether the behavioral response is activated. On the other hand, the anterior cingulate has previously been implicated in conflict monitoring, but not decision-making (Botvinick 2007). Thus, the activation of the anterior cingulate might suggest a conflict between the long-term negative consequences of excessive online gaming and the rewarding short-term pleasures of gaming.

According to these results, we hypothesized that, when subjects with IGD are exposed to a visual gaming cue, the precuneus participates in the processing of visual stimuli and in the recollection of positive experiences such as winning. The parahippocampus stores memories of the gaming experience and processes their emotional significance. The anterior cingulate then assesses the desire provoked by the motivation system and determines whether the gaming urge is activated. Lastly, the DLPFC is activated to choose and plan the online gaming behavior. However, this hypothesis requires further testing in future studies. Specifically, functional connectivity should be investigated in these areas, or an effective connectivity study should be performed to test whether the above model is applicable for evaluating the brain response to gaming cues.

3.3.1.2 Other Task fMRI Studies

Event-related color-word fMRI Stroop task (Dong et al. 2012a, b)
Dong et al. (2012b) recruited 12 adults with IGD and 12 healthy controls. Subjects were classified into an internet addiction disorder (IAD) group if they had a score of 80 or higher on the self-report questionnaire developed by Young (1998) (and psychometrically tested by Widyanto and McMurran 2004). Subjects with IA also met the criteria of the Chinese Internet addiction test (Wang 2009). The control group was screened by psychiatric interviews.
All participants were asked to perform a color-word stroop task while undergoing an fMRI scan. Three target color words were arranged randomly in congruent or incongruent trials. The participants were asked to push a single button in response to the color of the presented word. The inter-stimuli interval was 600–1,400 ms (average, 1,000 ms). All participants who agreed to undergo scanning received a cash reward, and all were informed that they would receive an additional cash reward for participating in the task performance part of the study. The IAD and control groups showed no performance difference.

The difference in BOLD signal during the Stroop effect between IAD and control groups was calculated as [(IAD incongruent-IAD congruent)-(control incongruent-control congruent)] after correcting for multiple test scores at the whole brain level with a false discovery rate set to a threshold of $p < 0.05$. The IAD group revealed higher activity over the anterior and posterior cingulate compared to the control group. The results of this study suggest that adults with IAD have impaired inhibitory control and diminished efficiency of cognitive control. This result supported a previous EEG study of Dong and his colleagues (2010). They evaluated the response inhibition by event-related potentials during a Go/No go task among 12 IAD-afflicted and 12 normal university students. The IAD group exhibited lower NoGo-N2 amplitude, higher NoGo-P3 amplitude, and longer NoGo-P3 peak latency than the normal group. These results suggest that IAD students had lower activation in the conflict detection stage and had to engage in more cognitive effort to complete the inhibition task.

Task for reward sensitivity and loss sensitivity (Dong et al. 2011)
Dong et al. (2011) recruited 12 males with IGD and 12 healthy controls. Subjects were enrolled in the IGD group if they had been diagnosed with IGD based on a score of 80 or higher on the self-report questionnaire developed by Young to test for internet addiction (Widyanto and McMurran 2004). All subjects in the IGD group also met the criteria of the Chinese internet addiction test (Wang 2009). The control group was screened by psychiatric interviews in which they were required to complete a reality-simulated guessing task. The subjects need to choose one of two cards in 245 guessing trials. Depending on the color of the chosen card the subjects either win (red playing cards) or lose (black playing cards) 10 Dollars. Subjects were informed that, after completing the scanning, they would receive a cash reward for participating in the study.

In the win condition, the IGD group revealed higher activation of the orbitofrontal cortex than controls. Under the loss condition, the IGD group revealed lower activation of the anterior cingulate than controls. The explanation proposed by the authors was that, compared to controls, the IGD group had higher reward sensitivity but lower loss sensitivity.

Ball throwing animation task for disembodiment state (Kim et al. 2012)
Kim et al. (2012) recruited 17 adolescents with internet addiction and 17 healthy controls. The diagnoses of internet addiction were based on scores of 50 or higher on the self-report questionnaire used in the Korean Adolescent Internet Addiction Scale (Kim et al. 2003). This block-design study required the subjects to perform a ball-throwing task that reflected either self-agency about ball throwing (Agency condition) or the location of a ball (control conduction). Two further variables were introduced in the experiment: In the so-called fixed viewpoint block, the two conditions mentioned above (agency and control) were only shown from a fixed view (Corradi-Dell'acqua et al. 2008). In contrast, the change view block presented the agency and control conditions from different perspectives (such as from the first-person or third-person view). The interaction between agency condition and changed viewpoint revealed the disembodiment-related condition. In this study, the significance threshold was set to $p < 0.0005$ and $k > 10$.

The experimental results showed that subjects with internet addiction had higher brain activation of disembodiment near the left temporo-parieto-occipital junction and near the right parahippocampual area. The duration of internet use correlated with the activation over the posterior left middle temporal gyrus. Based on these results, the authors concluded that adolescents with internet addiction tend to manifest disembodiment-related activation of the brain and may lack the brain development needed for identity formation.

Summary of task fMRI studies
Due to the limited number of studies in the literature, any conclusions regarding cognitive function in IGD would be premature. In fact, the effect of IGD on cognitive function is controversial. Since most controlled substances are known to have damaging effects on the brain, a reasonable assumption is that they impair cognitive function. However, most online games exercise many specific cognitive

functions (Granic et al. 2014). Further, since perfect performance in online gaming requires good cognitive function, the hypothesis that online gaming produces a deficit in cognitive function is questionable. Further studies to compare the cognitive functions and behavioral characteristics of subjects with IGD and causal gamer are needed to clarify the role of cognitive functions in the process of addiction to online gaming.

3.3.2 Resting fMRI Study

A growing body of evidence shows that several neural circuits exhibit spontaneous activity at rest. These slow-frequency fluctuations are temporally correlated within spatially distinct but functionally related networks. Studies consistently show that, in healthy subjects, numerous networks in this resting-state functional connectivity represent specific patterns of synchronous activity (Rosazza and Minati 2011). Evaluation of resting-state functional connectivity provides an opportunity to characterize distributed circuit abnormalities in neuropsychiatric illnesses. For example, seed analyses of the amygdala, insula, and nucleus accumbens have revealed reduced functional connectivity in subjects with opioid dependence (Upadhyay et al. 2010). Representing the specific neurobiological network underlying reward, affective and cognitive processes in terms of functional connectivity may reveal possible mechanisms of addictive disorder (Sutherland et al. 2012). Until now, however, only three studies have discussed resting state functional connectivity in IA.

3.3.2.1 Regional Homogeneity of IA in Dong et al. (2012a)

Dong et al. (2012a) recruited 15 males diagnosed with IGD according to the Young internet addiction scale (Widyanto and McMurran 2004) and 14 healthy comparison subjects. The cut-off point in this study was 80. Regional homogeneity (ReHo) was evaluated in a resting state during an 8-min fMRI scan. The significance threshold was set to $p < 0.05$ FDR correction.

Comparisons with controls showed that the IGD subjects had higher ReHo in the brain stem, inferior parietal lobe, left posterior cerebellum, and left middle frontal gyrus. However, the IGD subjects had lower ReHo in the temporal, occipital, and parietal brain regions. Functional connectivity analyses of the seed region of the inferior parietal lobe showed no significant difference between the two groups.

3.3.2.2 Regional Homogeneity of IA in Liu et al. (2010)

Liu recruited 19 college students with IA and 19 controls. The diagnoses of IA were based on a version of the Young diagnostic questionnaire modified by Beard and Wolf (2001). Regional homogeneity was evaluated by 9-min fMRI scans with the subjects in a resting state. The resulting statistical map was set at a combined threshold of $p < 0.001$ and a minimum cluster size of 270 mm^3, which resulted in a corrected threshold of $p < 0.05$.

In the resting state, the IAD group showed more brain regions in ReHo compared to controls. The increased ReHo was distributed over the cerebellum, brainstem, right cingulate gyrus, bilateral parahippocampus, right frontal lobe, left superior frontal gyrus, left precuneus, right inferior temporal gyrus, left superior temporal gyrus and middle temporal gyrus. The ReHo was not decreased in the IAD group.

3.3.2.3 Inter-regional Connectivity in Adolescents with IA (Hong et al. 2013)

Hong et al. (2013) recruited 12 adolescents diagnosed with IA based on the Young internet addiction scale (Widyanto and McMurran 2004) and 11 healthy control subjects. Functional connectivity was evaluated by fMRI scans performed for 6 min 45 s with the subjects in a resting state. Compared to the control group, the IA group showed lower functional connectivity spanning a distributed network. Most of the impaired connections involved the subcortical brain region. No between-group differences were noted in the average clustering coefficient, the characteristic path length, or the small worldness ratio, an indicator of the extent to which the synchronization networks of cortical neurons exhibit the small-world topology. This result suggested that, in this group, IA was associated with a large and widespread decrease in functional connectivity in the cortico-striatal circuit.

However, a limitation of this study was that classification of subjects into the IA group and the control group was not based on psychiatric interviews. On the other hand, the mean score of 57.00 ± 17.39 on the Young internet addiction scale was relatively low. Further, the controls had a mean IQ of 109.63, which suggested that they were not representative of a normal population. Scanning time was also relatively short. Moreover, although many indicators were analyzed in this study, the p value for significance was not controlled for multiple comparisons. Finally, this study had a small sample size.

3.3.2.4 Default Model Network of IA (Ding et al. 2013)

Ding recruited 17 adolescents with IGD and 24 controls. The diagnoses of IGD were based on a modified version of the self-report Young diagnostic questionnaire developed by Beard and Wolf (2001). Default network resting-state functional connectivity was evaluated by 4,400-s fMRI scans with the subjects in a resting state.

The default network was evaluated by the functional connectivity to the posterior cingulate as defined by WFU-Pick Atlas (Maldjian et al. 2003). The significance threshold was set to $p < 0.05$ with AlphaSim correction. The subjects with IGD exhibited increased functional connectivity (FC) in the bilateral cerebellum, posterior lobe, and middle temporal gyrus but decreased FC in the bilateral inferior parietal lobe and right inferior temporal gyrus. Further, connectivity with PCC was positively correlated with CIAS score in the right precuneus, posterior cingulate, thalamus, caudate, nucleus accumbens, supplementary motor area and lingual gyrus. However, since the comparison results showed no overlap with the results of the correlation study, the author suggested that the alterations were partially consistent with those in subjects with substance use disorder.

The limitation of the study is that the IAD diagnoses were based solely on self-report questionnaires. Furthermore, the mean CIAS score in the IAD group was 64.59. Since the cut-off point for CIAS is 64 in adolescents (Ko et al. 2005b), the severity of the disorder in the IGD group was relatively mild. Lastly, the number of subjects in this study was small.

3.3.2.5 Summary of Resting fMRI Studies

Three studies have reported increased functional connectivity or regional homogeneity over the cerebellum (Ding et al. 2013; Dong et al. 2012; Liu et al. 2010). However, Hong et al. (2013) reported impaired connectivity in subjects with involvement of subcortical brain regions. The inconsistent results of these four studies may have resulted from different definitions of IA, different indicators of functional connectivity, and different ages and numbers of subjects. Further resting fMRI studies with clear definitions, adequate numbers of subjects, and clear hypotheses are needed to obtain additional data regarding functional connectivity in IAD. Further, since online gaming might also train the brain, some subjects with IGD could hypothetically show improved functional connectivity However, other impairments in functional connectivity might contribute to their lack of control over their internet use. Thus, detailed studies of specific networks are still needed.

3.3.3 Arterial Spin-Labeled Perfusion Magnetic Resonance Imaging

Perfusion provides oxygen and nutrients to tissues and is closely tied to brain function, which is an essential indicator of psychiatric disorder. Arterial spin label (ASL) perfusion MRI offers absolute quantification of cerebral blood flow (CBF). In ASL technique, arterial blood water is magnetically labeled using radiofrequency irradiation. The magnetically labeled arterial water decays with T1 relaxation. Since ASL MRI provides absolute quantification of CBF, which is coupled to regional neural activity, it can also be used to measure resting brain function independently of any specific sensorimotor of a cognitive task (Detre et al. 2009).

Feng et al. (2013) recruited 15 adolescents with IGD and 18 controls. The diagnoses of IGD were based on self-reported results on the Young diagnostic questionnaire, as modified by Beard and Wolf (2001). The subjects were arranged to undergo scanning with a 3T MRI scanner. Pseudocontinuous ASL perfusion images were collected by 3D fast spin echo acquisition with background suppression, 1,500 ms labeling, and a post-labeling delay of 1,500 ms. Multiple comparison corrections were performed with AlphaSim program at a combined threshold of $p < 0.05$ and a minimum cluster size of 54 voxels.

Compared with the control group, the IGD group showed higher CBF in the left inferior temporal lobe, left parahippocampal gyrus/amygdala, right medial frontal lobe/anterior cingulate, bilateral insula, right middle temporal gyrus, right precentral gyrus, left supplementary motor area, left cingulate, and right inferior parietal lobe. The CBF was decreased in the left middle temporal gyrus, left middle occipital gyrus, and right cingulate gyrus. The authors suggest that IGD is a behavioral addiction that may share similar neurobiological abnormalities with other addictive disorders.

3.4 Limitations and Controversial Issues in Previous fMRI Studies of IGD or IA

The major problem in fMRI studies of IGD or IA is the extreme heterogeneity of the subjects resulting from their widely varying internet and online gaming activities. If the aim of the study is identifying potential predictors of IA, a highly homogeneous group is needed. However, that sample selection is limited by the constraints of minimizing confounds, e.g. that all participants with IGD need to be addicted to the same game in a study. This might explain why all studies had a limited sample size. Further, if the aim of the study is to discuss sequelae of IA, a specific group with similar addiction experiences should be selected, which might explain why most research targets subjects with IGD for brain imaging studies. However, the wide variability in internet games may result in different consequences. Further, most subjects with IGD had a perfect performance in online gaming, which requires the presence of cognitive functions such as decision making, attention, concentration, and response inhibition, even though their daily life function in the real world was impaired. Since most cognitive skills used in games are similar to those used in cognitive tasks of researches, a diagnosis of impaired brain function in a cognitive task should be made cautiously until the deficit is further evaluated or confirmed by a behavioral assessment. Thus, a rational hypothesis based on clinical experience and on the literature is essential before performing a brain imaging study.

Compared to other event-related studies (Murphy and Garavan 2004), the sample sizes used in fMRI studies of IGD have been small. Recruiting greater numbers of subjects would provide more robust results and would help prevent type II error. Furthermore, most studies have focused on male subjects despite reports of gender differences in factors associated with IA (Ko et al. 2007, 2008; Yeh et al. 2008). Although evaluations of brain reactions specific to males or to females are clearly needed, most studies have focused on males because males have a higher risk of IGD. Further studies are needed to study hypothesized mechanisms specific to females.

Finally, previous studies have not applied consistent definitions of IGD and IA. Since the DSM-5 has proposed diagnostic criteria for IGD, future works should consistently define IGD or IA according to DSM-5 criteria.

3.5 Future Studies of Internet Gaming Disorder

The findings of this literature review show that further fMRI studies are needed to investigate the following issues:

1. According to a previous review of event-related studies (Murphy and Garavan 2004), each group should consider a minimum of 20 subjects.
2. Since the number of subjects is usually limited, psychiatric interviews are needed to identify IGD based on DSM-5 criteria (Association 2013) or on other validated diagnostic criteria (Ko et al. 2005a; 2009b; Tao et al. 2010).
3. Before beginning a study, a rational hypothesis must be made based on clinical experience or on previous epidemiological, neuro-cognitive, molecular, or neurobiological studies.
4. Future studies should consider mechanisms specific to females.
5. An integrated team including a clinical psychiatrist, a psychologist, an expert in fMRI physiology, a brain imaging technician, and an expert in imaging data analysis would improve the design and implementation of an fMRI study.
6. Advanced techniques such as arterial spin-labeled perfusion and advanced analyses such as psychological interaction analysis or dynamic caudal model could be used to explore and/or validate hypothesized mechanisms of IGD. However, such studies should only be put in place after a reasonable hypothesis is developed by an integrated team.
7. A hypothesis testing design that integrates psychopathology, prospective behavioral presentations, neurocognitive performance, neurophysiology, and brain imaging studies would help to achieve a comprehensive understanding of IGD.

3.6 Conclusion

Studies of gaming cue-induced reactivity have shown that the DLPFC and parahippocampus have similar roles in the gaming urge response. This suggests that the decision making and emotional context representation involved in the cue-induced gaming urge may also have roles in IGD. However, any conclusion regarding the mechanisms of IGD or IA would be premature due to the many limitations of previous studies. Future work should apply the diagnostic criteria used in the DSM-5 when recruiting subjects with IGD. Further fMRI studies with adequate sample sizes of at least 20 participants, reasonable hypotheses, effective designs, and precise data analyses by integrated research teams, are needed for further elucidation of mechanisms of IGD or IA.

References

American Psychiatric Association (2000) Diagnostic and statistical manual of mental disorders, 4th edn. APA, Washington DC

American Psychiatric Association (2013) Diagnostic and statistical manual of mental disorders, 5th edn. APA, Arlington

Beard KW, Wolf EM (2001) Modification in the proposed diagnostic criteria for internet addiction. CyberPsychol Behav 4:377–383

Bonson KR, Grant SJ, Contoreggi CS et al (2002) Neural systems and cue-induced cocaine craving. Neuropsychopharmacology 26:376–386

Botvinick MM (2007) Conflict monitoring and decision making: reconciling two perspectives on anterior cingulate function. Cogn Affect Behav Neurosci 7:356–366

Cavanna AE, Trimble MR (2006) The precuneus: a review of its functional anatomy and behavioural correlates. Brain 129:564–583

Chiamulera C (2005) Cue reactivity in nicotine and tobacco dependence: a "multiple-action" model of nicotine as a primary reinforcement and as an enhancer of the effects of smoking-associated stimuli. Brain Res Rev 48:74–97

Corradi-Dell'acqua C, Ueno K, Ogawa A et al (2008) Effects of shifting perspective of the self: an fMRI study. Neuroimage 40:1902–1911

Criaud M, Boulinguez P (2013) Have we been asking the right questions when assessing response inhibition in go/no-go tasks with fMRI? A meta-analysis and critical review. Neurosci Biobehav Rev 37:11–23

Davis RA (2001) A cognitive-behavioral model of pathological internet use. Comput Hum Behav 17:187–195

Detre JA, Wang J, Wang Z, Rao H (2009) Arterial spin-labeled perfusion MRI in basic and clinical neuroscience. Curr Opin Neurol 22:348–355

Ding WN, Sun JH, Sun YW et al (2013) Altered default network resting-state functional connectivity in adolescents with internet gaming addiction. PLoS One 8:e59902

Dong G, Zhou H, Zhao X (2010) Impulse inhibition in people with internet addiction disorder: electrophysiological evidence from a Go/NoGo study. Neurosci Lett 485:138–142

Dong G, Huang J, Du X (2011) Enhanced reward sensitivity and decreased loss sensitivity in internet addicts: an fMRI study during a guessing task. J Psychiatr Res 45:1525–1529

Dong G, Huang J, Du X (2012a) Alterations in regional homogeneity of resting-state brain activity in internet gaming addicts. Behav Brain Funct 8:41

Dong G, Devito EE, Du X, Cui Z (2012b) Impaired inhibitory control in 'internet addiction disorder': a functional magnetic resonance imaging study. Psychiatry Res Neuroimaging 203:153–158

Feng Q, Chen X, Sun J et al (2013) Voxel-level comparison of arterial spin-labeled perfusion magnetic resonance imaging in adolescents with internet gaming addiction. Behav Brain Funct 9:33

Fowler JS, Volkow ND, Kassed CA, Chang L (2007) Imaging the addicted human brain. Sci Pract Perspect 3:4–16

Friston KJ, Penny WD, Glaser DE (2005) Conjunction revisited. Neuroimage 25:661–667

Glover GH (2011) Overview of functional magnetic resonance imaging. Neurosurg Clin N Am 22:133–139

Goldman-Rakic P, Leung HC (2002) Functional architecture of the dorsolateral prefrontal cortex in monkeys and humans, in principles of frontal lobe function. Oxford University Press, Oxford

Granic I, Lobel A, Engels RC (2014) The benefits of playing video games. Am Psychol 69:66–78

Han DH, Hwang JW, Renshaw PF (2010) Bupropion sustained release treatment decreases craving for video games and cue-induced brain activity in patients with internet video game addiction. Exp Clin Psychopharmacol 18:297–304

Heinz A, Beck A, Grusser SM et al (2009) Identifying the neural circuitry of alcohol craving and relapse vulnerability. Addict Biol 14:108–118

Hong SB, Zalesky A, Cocchi L et al (2013) Decreased functional brain connectivity in adolescents with internet addiction. PLoS One 8:e57831

Huettel SA, Song AW, McCarthy G (2009) Functional magnetic resonance imaging, 2nd edn. Sinauer, Massachusetts

Kilts CD, Schweitzer JB, Quinn CK et al (2001) Neural activity related to drug craving in cocaine addiction. Arch Gen Psychiatry 58:334–341

Kim EJ, Lee SY, Oh SG (2003) The validation of Korean adolescent internet addiction scale (K-AIAS). Korean J Clin Psychol 22:5

Kim YR, Son JW, Lee SI et al (2012) Abnormal brain activation of adolescent internet addict in a ball-throwing animation task: possible neural correlates of disembodiment revealed by fMRI. Prog Neuropsychopharmacol Biol Psychiatry 39:88–95

Ko CH, Yen JY, Chen CC et al (2005a) Proposed diagnostic criteria of internet addiction for adolescents. J Nerv Ment Dis 193:728–733

Ko CH, Yen JY, Yen CF et al (2005b) Screening for internet addiction: an empirical study on cut-off points for the chen internet addiction scale. Kaohsiung J Med Sci 21:545–551

Ko CH, Yen JY, Chen CC et al (2006) Tridimensional personality of adolescents with internet addiction and substance use experience. Can J Psychiatry 51:887–894

Ko CH, Yen JY, Yen CF et al (2007) Factors predictive for incidence and remission of internet addiction in young adolescents: a prospective study. CyberPsychol Behav 10:545–551

Ko CH, Yen JY, Yen CF et al (2008) The association between internet addiction and belief of frustration intolerance: the gender difference. CyberPsychol Behav 11:273–278

Ko CH, Liu GC, Hsiao S et al (2009a) Brain activities associated with gaming urge of online gaming addiction. J Psychiatry Res 43:739–747

Ko CH, Yen JY, Chen SH et al (2009b) Proposed diagnostic criteria and the screening and diagnosing tool of internet addiction in college students. Compr Psychiatry 50:378–384

Ko CH, Yen JY, Yen CF et al (2012) The association between internet addiction and psychiatric disorder: a review of the literature. Eur Psychiatry 27:1–8

Ko CH, Liu GC, Yen JY et al (2013a) Brain correlates of craving for online gaming under cue exposure in subjects with internet gaming addiction and in remitted subjects. Addict Biol 18:559–569

Ko CH, Liu GC, Yen JY et al (2013b) The brain activations for both cue-induced gaming urge and smoking craving among subjects comorbid with internet gaming addiction and nicotine dependence. J Psychiatry Res 47:486–493

Liu J, Gao XP, Osunde I et al (2010) Increased regional homogeneity in internet addiction disorder: a resting state functional magnetic resonance imaging study. Chin Med J 123:1904–1908

Lorenz RC, Kruger JK, Neumann B et al (2013) Cue reactivity and its inhibition in pathological computer game players. Addict Biol 18:134–146

Maldjian JA, Laurienti PJ, Kraft RA, Burdette JH (2003) An automated method for neuroanatomic and cytoarchitectonic atlas-based interrogation of fMRI data sets. Neuroimage 19:1233–1239

Montag C, Weber B, Trautner P et al (2012) Does excessive play of violent first-person-shooter-video-games dampen brain activity in response to emotional stimuli? Biol Psychol 89:107–111

Montag C, Bey K, Sha P et al (2014) Is it meaningful to distinguish between generalized and specific Internet addiction? Evidence from a cross-cultural study from Germany, Sweden, Taiwan and China. Asia-Pacific Psychiatry
Mitchell DG (2011) The nexus between decision making and emotion regulation: a review of convergent neurocognitive substrates. Behav Brain Res 217:215–231
Murphy K, Garavan H (2004) An empirical investigation into the number of subjects required for an event-related fMRI study. Neuroimage 22:879–885
Nichols T, Brett M, Andersson J et al (2005) Valid conjunction inference with the minimum statistic. Neuroimage 25:653–660
Risinger RC, Salmeron BJ, Ross TJ et al (2005) Neural correlates of high and craving during cocaine self-administration using BOLD fMRI. Neuroimage 26:1097–1108
Rosazza C, Minati L (2011) Resting-state brain networks: literature review and clinical applications. Neurol Sci 32:773–785
Rudy JW (2009) Context representations, context functions, and the parahippocampal-hippocampal system. Learn Mem 16:573–585
Salzmann E, Vidyasagar TR, Creutzfeldt OD (1993) Functional comparison of neuronal properties in the primate posterior hippocampus and parahippocampus (area TF/TH) during different behavioural paradigms involving memory and selective attention. Behav Brain Res 53:133–149
Skinner MD, Aubin HJ (2010) Craving's place in addiction theory: contributions of the major models. Neurosci Biobehav Rev 34:606–623
Sutherland MT, McHugh MJ, Pariyadath V, Stein EA (2012) Resting state functional connectivity in addiction: lessons learned and a road ahead. Neuroimage 62:2281–2295
Tao R, Huang X, Wang J et al (2010) Proposed diagnostic criteria for internet addiction. Addiction 105:556–564
Tejado Lde A, Ruiz RM, Trebbau H et al (2010) Functional magnetic resonance studies in eating behavior disorders. Actas Esp Psiquiatry 38:183–188
Upadhyay J, Maleki N, Potter J et al (2010) Alterations in brain structure and functional connectivity in prescription opioid-dependent patients. Brain 133:2098–2114
Volkow ND, Wang GJ, Fowler JS et al (2010) Addiction: decreased reward sensitivity and increased expectation sensitivity conspire to overwhelm the brain's control circuit. BioEssays 32:748–755
Wang W, Tao R, Niu Y et al (2009) Preliminarily proposed diagnostic criteria of pathological internet use. Chin Ment Health J 23:5
Widyanto L, McMurran M (2004) The psychometric properties of the internet addiction test. CyberPsychol Behav 7:443–450
Yeh YC, Ko HC, Wu JY, Cheng CP (2008) Gender differences in relationships of actual and virtual social support to internet addiction mediated through depressive symptoms among college students in Taiwan. CyberPsychol Behav 11:485–487
Young KS (1998) Caught in the net: how to recognize the signs of internet addiction and a winning strategy for recovery. Wiley, New York

Chapter 4
Internet Addiction and PET

Hyun Soo Park and Sang Eun Kim

Abstract Pathological use of the Internet is a new and rapidly growing worldwide phenomenon; however, few studies have examined the neurobiological factors underlying this condition. Internet addiction has been considered a behavioral addiction that is accompanied by withdrawal symptoms and tolerance, characteristics that may result from abnormalities in neural substrates involved in impulse control and reward processing. Recent studies on Internet addiction have highlighted symptoms of cognitive and emotional dysfunction that are similar to other types of drug and/or behavioral addiction. In this chapter, we describe the results of neurobiological investigations of the underlying mechanisms of Internet addiction using positron emission tomography (PET). The altered cerebral glucose metabolism and the reduced striatal dopamine D2 receptor availability found in people exhibiting pathological use of the Internet are discussed in terms of the similarities of these characteristics to those observed in substance abusers.

4.1 Introduction

Everyone is using the Internet these days, it seems—all the time, and everywhere. Internet addiction is a new, rapidly growing and worrying addictive worldwide phenomenon (Young 2010). Extreme cases of Internet addiction have even

H.S. Park · S.E. Kim (✉)
Department of Nuclear Medicine, Seoul National University Bundang Hospital, Seoul National University College of Medicine, Seoul, Korea
e-mail: kse@snu.ac.kr

H.S. Park
e-mail: hyuns@snu.ac.kr

H.S. Park · S.E. Kim
Graduate School of Convergence Science and Technology, Seoul National University, Seoul, Korea

S.E. Kim
Advanced Institutes of Convergence Technology, Seoul, Korea

C. Montag and M. Reuter (eds.), *Internet Addiction*, Studies in Neuroscience, Psychology and Behavioral Economics, DOI 10.1007/978-3-319-07242-5_4

been reported with very dire consequences (CNN 2010). More than one out of eight adults in the United States shows signs of being addicted to the Internet (Aboujaoude et al. 2006). A similar or higher rate has been reported in several countries in Asia (Yen et al. 2007) and Europe (Ferraro et al. 2007). Among these, South Korea may be one of the countries that suffer the most from this phenomenon. In 2011, the National Information Society Agency of Korea reported that 6.8 % of Korean adults were classified as exhibiting a definitive Internet addiction, defined as using the Internet for an average of 2.7 h a day. It was noted that 41.3 % of the Internet addicts' online time was spent playing interactive online games (Ko et al. 2012). For many researchers, Internet addiction has been considered seriously as a type of behavioral addiction that is accompanied by withdrawal symptoms, tolerance and comorbid psychiatric symptoms (mood disorders, depression, etc.) that contribute to the pathological pattern of behavior. However, despite the increasing number of individuals suffering from Internet addiction, few studies have examined the neurobiological factors underlying this condition.

In this chapter, we introduce the results of PET studies on Internet addiction using radioligands designed to assess glucose metabolism [^{18}F-2-fluorodeoxyglucose (FDG)] and dopamine availability [^{11}C-raclopride (RAC)]. Specifically, we examined the differences in neural substrates and dopaminergic function between young individuals with Internet addiction and those with normal use. We hypothesized that the Internet addicts would show altered cerebral glucose metabolism in the prefrontal and striatal regions, which are areas of the brain implicated in impulse control and reward processing. In addition, we expected that the addicts would exhibit abnormal dopamine D2 receptor availability in the striatal brain regions, given that these receptor populations are known to modulate reinforcement in addiction.

4.2 PET Studies of Neurobiological Mechanisms of Addiction

PET is designed to map biological and physiological processes in living subjects following administration of positron-emitting radioisotopes. This nuclear imaging technique is based on the detection of photons released by annihilation of positrons emitted by radioisotopes. Positron-emitting radioisotopes are produced in a cyclotron by bombarding target material with accelerated protons. When such a radioisotope is injected into the body of a living subject, the radionuclides emit positrons that are annihilated with nearby electrons, resulting in the release of two photons. PET detects these events and translates this information to reveal the location and extent (radioactive concentration) of the annihilation. The resulting image data can be used to determine the distribution of radioligand in the body (Miyoshi et al. 2011). PET provides an opportunity to observe biological and physiological phenomena in vivo. The application of PET for neurobiological studies can be variously extended when a radioisotope, such as ^{18}F and ^{11}C, is labeled with molecules that exhibit agonism or antagonism for the various receptors or enzymes. In this sense, the radiolabeled molecules act like *tracers* for the target substrates and PET monitoring of the behavior of the tracer reveals the status of the target receptors (e.g., neurotransmitter availability) or metabolism (e.g., glucose consumption). By using this concept, for example, not only can we measure altered glucose metabolism related to an abnormal condition of tissue cells using a glucose analogue labeled with ^{18}F (^{18}F-FDG), but we can also assess pathological neurotransmitter availability underlying psychiatric disorders using 1 radiotracers for specific types of neurotransmission. Because these measurements are noninvasive, the technology allows researchers to track biochemical transformation in the living human and animal body without perturbing the system that is measured. For decades, functional neuroimaging using PET and radioligands has gained widespread acceptance as a tool of psychiatric research. Because ^{18}F-FDG PET images represent activation of regional brain function, researchers frequently use this technique to map differences between normal and abnormal regional brain function.

Abnormal cerebral activation linked with addiction is primarily found in the frontal cortex, especially in the pre- and/or orbitofrontal cortex, with both hypo- and hyper-metabolism being reported. PET studies have shown that cocaine (Volkow et al. 1993) and methamphetamine (Volkow et al. 2001; Bolla et al. 2003) reduce the activity of the orbitofrontal cortex, an area which is implicated in executive functions, such as decision-making, planning and judgment; notably, these functions contribute to the inhibition of impulsive behavior in normal subjects. Poor judgment has in fact been associated with lower activation of the orbitofrontal cortex in cocaine addicts (Bolla et al. 2003). Both cocaine and alcohol addicts showed increased relative activation of the orbitofrontal cortex which was associated with improved performance of a cognitive task; by comparison, normal control subjects showed worse performance with increased orbitofrontal activation, suggesting that a reversal of the role of the orbitofrontal cortex occurs as a function of addiction (Goldstein et al. 2001). Orbitofrontal activation was significantly and dose-dependently decreased and executive function was reduced in the group of methamphetamine abusers compared to control subjects (Kim et al. 2009). The relationship between orbitofrontal activation (hypo- and hyper-metabolism) and performance of cognitive task seem contradictory. Greater activation in the orbitofrontal cortex may therefore be indicative of task-independent mental hyperactivity and the greater effort required to implement inhibitory control. That is, Internet game addicts may be more likely to be engaged in random thoughts and to constantly place them in regulatory control mode for better everyday performance. Alternatively, it is possible that the association between higher activity in the orbitofrontal cortex and impulsivity is nonlinear; if this was the case, either high activation or low activation might result in abnormal impulse control.

Abnormal activation in the orbitofrontal cortex has also been observed in individuals with a behavioral addiction, such as pathological gambling. Hollander and colleagues demonstrated that relative glucose metabolic rates in the orbitofrontal cortex and medial frontal cortex were significantly increased at baseline in a group of pathological gamblers compared to normal controls (Hollander et al. 2008). We can hypothesize that such deficits in metabolism in the orbitofrontal cortex may result in the improper inhibitory control and compulsive behaviors which comprise the cognitive behavioral aspects of addiction.

PET has shown its greatest value in allowing psychiatric researchers to better understand the relationships between neurotransmitter systems and psychiatric diseases, including substance abuse disorder. Much of this work has focused on the dopamine system, contributing to the supposition that dopamine plays a key role in addiction (Volkow et al. 1993, 2002, 2009; Wang et al. 2001). A deteriorated reward processing function associated with reduced dopamine D2 receptor availability primarily in the striatum, a complex of the caudate nucleus and the putamen, has been reported repeatedly in addicted individuals (Volkow et al. 1990, 1993, 1996, 2001; Wang et al. 1997). This is also supported by evidence showing that A1+ carriers of the ANKK1/DRD2 Taq Ia polymorphisms are prone to get addicted to the Internet and other addictions such as to alcoholism (Munafo et al. 2007). Of note, the A1 allele is associated with a 30–40 % reduced dopamine 2 receptor density in striatal regions (Noble 2000).

A group has conducted a set of PET studies using multiple tracers across studies to investigate the relationship between dopaminergic function and brain glucose metabolism in the prefrontal cortex. These studies demonstrated the presence of negative association between brain glucose metabolism in prefrontal cortical regions and changes in dopamine levels by methylphenidate, a dopamine transporter blocker, treatment in the striatum of control subjects. Thus, the higher the metabolism in the prefrontal region, the lower the methylphenidate-induced changes in striatal dopamine levels. In contrast, glucose metabolism in the prefrontal cortical regions was not correlated with striatal dopamine changes in alcoholics (Volkow et al. 2007). These findings suggest that in addicts the normal regulation of striatal dopamine activity by signals from the prefrontal cortex is disrupted; thus, the decreased striatal dopamine activity of addicts may reflect abnormal prefrontal regulation of the mesolimbic dopamine system (Thanos et al. 2008). Dopamine is believed to play the most significant role in mediating drugs' reinforcing effects by acting on the mesolimbic dopamine system. Furthermore, accumulating evidence has suggested that abnormalities in the dopaminergic system play an important role in various types of behavioral addiction that do not involve chemical substances (Holden 2001).

4.3 Altered Regional Glucose Metabolism in Internet Game Addicts

Recent research has suggested that Internet addiction may be based on psychological and cognitive mechanisms that are similar to those underlying pathological gambling (Johansson and Gotestam 2004; Goudriaan et al. 2005; Hollander et al. 2005; Aboujaoude et al. 2006). According to the Diagnostic and Statistical Manual of Mental Disorders (DSM), pathological gambling is a type of impulse control disorder that involves the inability to resist the impulse to perform an action that is harmful to oneself or others. Internet addiction is similar to pathological gambling as it is does not directly involve intoxicant or psychoactive substances, but entails dysfunctions in impulse control (Shapira et al. 2000) and reward processing (Han et al. 2007; Ko et al. 2010). Thus, Internet addiction may result from abnormalities in neural substrates that are involved in impulse control and reward processing. Based on the previous results, investigators have established an opinion that Internet addiction should be included in DSM as an impulse control disorder or as a definitive disorder. Recently, Internet Gaming Disorder was mentioned in Section III as a condition warranting more clinical research and experience before it might be considered for inclusion in DSM as a formal disorder.

To investigate differences in cerebral activation (regional cerebral glucose metabolism) between individuals with pathological Internet game use and those with normal use, we performed a ^{18}F-FDG PET study. We hypothesized that the addicts would show altered metabolism in the orbitofrontal and striatal regions, which are implicated in impulse control and reward processing, respectively.

Severity of addictive Internet game use was assessed using a standardized questionnaire for Internet game addiction (IGS) (Oh et al. 2005) and impulsiveness was assessed using the Barratt Impulsiveness Scale Version 11 (BIS) (Patton et al. 1995). An interview regarding game use was also conducted. Among 464 respondents who received the questionnaire, 7.9 % were identified as Internet addicts who habitually engage in game playing. The score of the IGS was a sum of level of game-dependent behavior, declined self-control, hypersensitivity, functional deficits and absorption in Internet game use; naturally, scores were higher in the identified group of Internet addicts than in the normal user group. Participants in the addict group reportedly spent an average of 22.6 h per week on Internet games during the previous month and 63.6 % had been playing Internet games >4 years. The longest continuous span during which the addicts had played a game without a break was 11.5 h on average. The most common type of game was a multiplayer, online strategy simulation game (e.g., World of Warcraft III and Starcraft II, Blizzard Entertainment, Inc.). Consistent with previous studies investigating impulsive traits of addicts, results of the BIS score showed that the game addicts were more impulsive than normal users (Fig. 4.1a). Moreover, IGS scores were positively correlated with level of impulsiveness in the addicted group (Fig. 4.1b).

Significant differences in cerebral regional activation (resting-state glucose metabolism) between the Internet game addicts and the normal users were demonstrated using ^{18}F-FDG PET. Participants had not been engaged in any psychological and/or pharmacological intervention during the ^{18}F-FDG PET study. The Internet game addicts had significantly increased activation in the right middle orbitofrontal gyrus (BA 11), the left caudate nucleus, and the right insula (BA 13), compared to the normal users (Fig. 4.2). In addition, the Internet game addicts had significantly decreased activation in the bilateral postcentral gyrus (BA 2/3), the left precentral gyrus (BA 4), and the right superior parietal lobule (BA 7), as well as in the right superior occipital gyrus (BA 18) and the left inferior occipital gyrus (BA 19).

We found that Internet game addicts had greater activation in the orbitofrontal cortex, striatum, and somatosensory regions, which are areas of the brain implicated in impulse control, reward processing, and somatic representation.

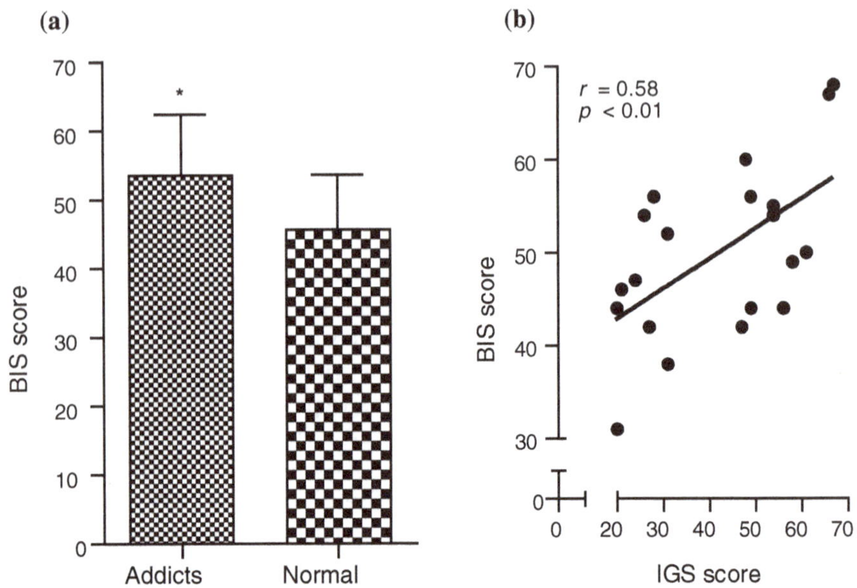

Fig. 4.1 **a** Comparison of BIS scores between the Internet game addicts and normal users, **b** relationship between the severity of Internet game addiction (IGS score) and the level of impulsiveness (BIS score), *BIS* barratt impulsiveness scale, *IGS* Internet game addiction scale (Park et al. 2010). $^*p < 0.05$

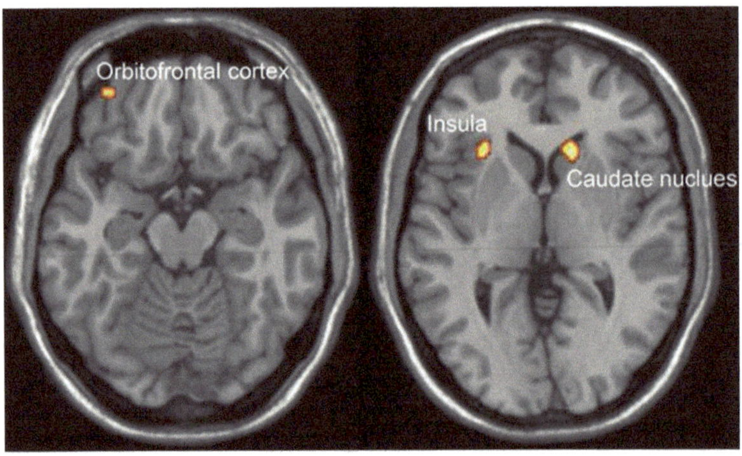

Fig. 4.2 A representative image of increased activation in Internet game addicts (Park et al. 2010)

The orbitofrontal cortex has been implicated in impulsivity and inhibitory control (Buckholtz et al. 2010). For example, individuals with borderline personality disorder, who suffer from high levels of impulsivity, have shown significantly reduced glucose metabolism in the medial orbitofrontal cortex (Soloff et al. 2003).

In substance abusers, higher glucose metabolism in the orbitofrontal cortex has been associated with better performance on the Stroop task, a measure of the ability to inhibit impulses, or more specifically, a prepotent response tendency (Goldstein et al. 2001). Bolla et al. 2003 found that better performance on a computerized risk-taking task was associated with greater activation in the right orbitofrontal cortex during the task in both cocaine addicts and normal controls. These previous studies suggested that abnormal activity (either higher or lower activation) in the orbitofrontal cortex was associated with dysfunctional impulsive behavior and response control.

The striatum has been strongly implicated in addiction and reward processing (Robbins and Everitt 1999; Hyman and Malenka 2001; Volkow et al. 2002). In the present study, the Internet game addicts had increased activity in the left caudate nucleus. These results are consistent with the findings of previously reported studies. For example, methamphetamine abusers had increased activity in brain regions including the ventral striatum and lateral orbitofrontal cortex (London et al. 2004). Moreover, various types of substance abusers and behavioral addicts showed decreased dopamine D2 receptor availability (Volkow et al. 2008; Pallanti et al. 2010). However, a functional magnetic resonance imaging (fMRI) study showed a reduction of ventral striatal and ventromedial prefrontal activation during a gambling task in pathological gamblers, which was correlated with gambling severity (Reuter et al. 2005). The striatal functional activity associated with substance and non-substance related addiction is unclear. Because the striatum and orbitofrontal cortex receive extensive dopaminergic innervation (Volkow and Fowler 2000; Volkow et al. 2002; Hollander et al. 2005; Chang and Haning 2006), altered metabolic activities in these regions may be associated with altered dopaminergic neurotransmission in Internet game addicts, as suggested by a genetic study (Han et al. 2007).

The insula is known to play a crucial part in addiction because of its role in conscious urges to take drugs of abuse. Recently, an fMRI study revealed a cue-induced activation of the right insula in subjects with Internet game addiction (Ko et al. 2009). These results, together with our data, support the notion that the insula is involved in conscious urges to engage in addictive behaviors and the decision-making processes that precipitate relapse.

This study demonstrated that Internet game addiction may be associated with abnormal neurobiological mechanisms in the orbitofrontal cortex, striatum, and somatosensory regions, which are implicated in impulse control, reward processing, and somatic representation of previous experiences. Altered resting glucose metabolism in the ventromedial prefrontal cortex and striatum may be associated with abnormal dopaminergic neurotransmission in Internet game addiction similarly to associations that have been observed in subjects with substance abuse and other types of behavioral addiction. Together with previous behavioral and cognitive findings on Internet game addiction, our imaging and behavioral results suggest that Internet game addiction may be considered a type of impulse control disorder or non-substance-related addiction that shares psychological and neural mechanisms with substance-related addiction.

4.4 Reduced Striatal Dopamine D2 Receptor Availability in Internet Addicts

Abnormalities in the dopaminergic neural system have been widely reported in individuals with substance abuse disorders. For example, reductions in dopamine D2 receptors and in dopamine release have been consistently observed in individuals who are addicted to cocaine, marijuana, or alcohol in studies using PET and the radiolabeled dopamine D2 receptor antagonist ^{11}C-RAC (Volkow et al. 2009). Moreover, accumulating evidence has suggested that abnormalities in the dopaminergic system may also play an important role in various types of behavioral addictions that do not involve chemical substances (Holden 2001). For example, a recent ^{18}F-FDG PET study found that patients who exhibit pathological gambling have altered levels of glucose metabolism in the ventral parts of the striatum compared to normal controls (Pallanti et al. 2010). People with morbid overeating behavior were also found to have lower densities of dopamine receptors in the striatum (Wang et al. 2001; Volkow et al. 2008). Taken together, these findings suggest that behavioral addiction, like substance-related addiction, may be partly due to impaired dopaminergic neural systems (Reuter et al. 2005; Potenza 2006). Given the evidence that Internet addiction shares neurocognitive and emotional deficits with other types of addiction (Park et al. 2010), we hypothesized that people with Internet addiction would have reduced dopaminergic receptor availability in striatal brain regions. To test this hypothesis, we recruited adult men with or without Internet addiction. Although the Internet addiction is thought to be categorized into 'generalized' and 'specific' according to the domain·(video gaming, shopping, social networking and/or pornography) of problematic use of Internet (Davis 2001; Montag et al. 2014), we didn't distinguish participant's type of Internet addiction in the present study. Internet addiction was assessed using Young's Internet Addiction Test (IAT) questionnaire (Young 1998). Dopamine D2 receptor availability was assessed with PET and the radioligand ^{11}C-RAC, which binds to dopamine D2 receptors without eliciting neurochemical response.

We found that Internet addicts have significantly reduced dopamine D2 receptor availability in the left dorsal caudate and putamen, and in the right dorsal caudate. A trend for reduced dopamine D2 receptor availability in the addict group was also seen in the right putamen (Fig. 4.3a). Moreover, the relationship between dopamine D2 receptor availability and Internet addiction score was significant in the left dorsal caudate and putamen (Fig. 4.3b).

These findings suggest that Internet addiction is associated with dysfunctions in the brain dopaminergic systems, consistent with previous reports of reduced dopamine D2 receptor availability in both substance-related and behavioral addiction (Volkow et al. 2008; Pallanti et al. 2010). The striatum is strongly implicated in addiction, reward processing, and reinforcement. In the present studies, addicts had increased glucose metabolism in the left caudate nucleus and bilaterally decreased dopamine D2 receptor availability in caudate nucleus and putamen. These results are consistent with the findings of previously reported studies. For example,

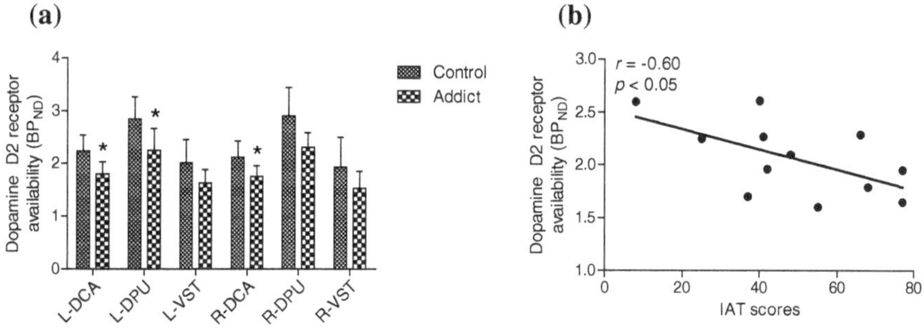

Fig. 4.3 **a** Comparison of dopamine D2 receptor availability in subdivisions of the striatum between Internet addicts and normal controls, and **b** the relationship between the severity of Internet addiction (IAT score) and dopamine D2 receptor availability in the left dorsal caudate *L* left; *R* right; *DCA* dorsal caudate; *DPU* dorsal putamen; *VST* ventral striatum; *IAT* Internet addiction test (Kim et al. 2011). *$p < 0.05$

methamphetamine abusers had increased glucose metabolism in brain regions including the ventral striatum and lateral orbitofrontal cortex (London et al. 2004). Moreover, numerous studies have reported that decreased striatal dopamine D2 receptor availability is associated with various addictive states (Volkow et al. 2008; Pallanti et al. 2010). Reduced dopamine receptor availability has been suggested to contribute to reward deficiency syndrome, characterized by the presence of an increased need for high levels of excitement and stimulation in people with genetically derived alterations in dopaminergic neurotransmission in the reward pathway (Comings and Blum 2000). People who suffer from this syndrome may seek drugs or alternatives to drugs such as gambling, in order to normalize their hypodopaminergic activity. Using Internet games and being engaged in online activities may also stimulate the reward pathway, consequently eliciting feelings of excitement and pleasure. In fact, experiencing pleasure was found to be the strongest predictor of Internet addiction (Chou and Hsiao 2000). Recent neuroimaging studies have also provided supporting evidence that Internet activities, such as online games, stimulated dopaminergic brain areas including the striatum and prefrontal regions (Ko et al. 2009; Han et al. 2010). In an fMRI study, people who were addicted to the popular online game "World of Warcraft" elicited greater activation in brain regions including the striatum, orbitofrontal, and dorsolateral prefrontal regions when viewing visual images from the game compared to non-addicted controls (Ko et al. 2009). More direct evidence suggesting increased dopamine release during Internet gaming can be drawn from a neurochemical imaging study in which research participants were asked to play a video game that involved navigating a tank for monetary reward. In this study, dopamine D2 receptor availability in the dorsal and ventral striatum were decreased in participants playing the video game compared with a control activity of viewing a black screen (Koepp et al. 1998). Therefore, currently available neuroimaging evidence suggests that the Internet may provide a rewarding experience and that Internet addiction is possibly associated with lower levels of dopaminergic activity in brain reward pathways, similar to other addictive disorders. Thus, our findings support the claim that Internet addiction may be characterized by neurobiological abnormalities similar to those of other addictive disorders (Potenza 2006; Park et al. 2010).

4.5 Conclusion

Our data demonstrate that Internet gaming addicts show abnormal resting-state glucose metabolism in the orbitofrontal cortex, striatum, and insular cortex, which are regions implicated in impulse control, reward processing, and modulating urges. In addition, Internet-addicted subjects exhibited bilaterally decreased dopamine D2 receptor availability in the dorsal striatum, an area of the brain implicated in the modulation of reinforcement in addiction. This finding is consistent with previous results and contributes to our understanding of the neurobiological mechanisms of Internet addiction. Our results suggest that Internet game addiction should be considered an impulse control disorder or non-substance related addiction that shares psychological and neural mechanisms with substance-related addictions.

References

Aboujaoude E, Koran L, Gamel N et al (2006) Potential markers for problematic internet use: a telephone survey of 2,513 adults. CNS Spectr 11:750–755

Bolla KI, Eldreth DA, London ED et al (2003) Orbitofrontal cortex dysfunction in abstinent cocaine abusers performing a decision-making task. Neuroimage 19:1085–1094

Buckholtz JW, Treadway MT, Cowan RL et al (2010) Dopaminergic network differences in human impulsivity. Science 329:532

Chang L, Haning W (2006) Insights from recent positron emission tomographic studies of drug abuse and dependence. Curr Opin Psychiatry 19:246–252

CNN (2010) Report: South Korean couple starved child while raising 'virtual baby'. Available via http://edition.cnn.com/2010/WORLD/asiapcf/03/05/korea.baby.starved/

Chou C, Hsiao MC (2000) Internet addiction, usage, gratification, and pleasure experience: the Taiwan college students' case. Comput Educ 35:65–80

Comings DE, Blum K (2000) Reward deficiency syndrome: genetic aspects of behavioral disorders. Prog Brain Res 126:325–341

Davis RA (2001) A cognitive-behavioral model of pathological internet use. Comput Hum Behav 17:187–195

Ferraro G, Caci B, D'Amico A, Di Blasi M (2007) Internet addiction disorder: an Italian study. Cyberpsychol Behav 10:170–175

Goldstein RZ, Volkow ND, Wang GJ et al (2001) Addiction changes orbitofrontal gyrus function: involvement in response inhibition. NeuroReport 12:2595–2599

Goudriaan AE, Oosterlaan J, de Beurs E, van den Brink W (2005) Decision making in pathological gambling: a comparison between pathological gamblers, alcohol dependents, persons with Tourette syndrome, and normal controls. Brain Res Cogn Brain Res 23:137–151

Han DH, Lee YS, Yang KC et al (2007) Dopamine genes and reward dependence in adolescents with excessive internet video game play. J Addict Med 1:133–138

Han DH, Kim YS, Lee YS et al (2010) Changes in cue-induced, prefrontal cortex activity with video-game play. Cyberpsychol Behav Soc Netw 13:655–661

Hollander E, Buchsbaum MS, Haznedar MM et al (2008) FDG-PET study in pathological gamblers. 1. Lithium increases orbitofrontal, dorsolateral and cingulate metabolism. Neuropsychobiology 58:37–47

Holden C (2001) 'Behavioral' addictions: do they exist? Science 294:980–982

Hollander E, Pallanti S, Baldini Rossi N et al. (2005) Imaging monetary reward in pathological gamblers. World J Biol Psychiatry 6:113–120

Hyman SE, Malenka RC (2001) Addiction and the brain: the neurobiology of compulsion and its persistence. Nat Rev Neurosci 2:695–703

Johansson A, Gotestam KG (2004) Problems with computer games without monetary reward: similarity to pathological gambling. Psychol Rep 95:641–650

Kim YT, Lee SW, Kwon DH et al (2009) Dose-dependent frontal hypometabolism on FDG-PET in methamphetamine abusers. J Psychiatr Res 43:1166–1170

Kim SH, Baik SH, Park CS et al (2011) Reduced striatal dopamine D2 receptors in people with internet addiction. NeuroReport 22:407–411

Ko CH, Hsiao S, Liu GC et al (2010) The characteristics of decision making, potential to take risks, and personality of college students with internet addiction. Psychiatry Res 175:121–125

Ko CH, Liu GC, Hsiao S et al (2009) Brain activities associated with gaming urge of online gaming addiction. J Psychiatr Res 43:739–747

Ko YS, Uhm NR, Jeon JS (2012) A survey of internet addiction (NIA IV-RER-11052). National Information Society Agency, Ministry of Public Administration and Security, Seoul

Koepp MJ, Gunn RN, Lawrence AD et al (1998) Evidence for striatal dopamine release during a video game. Nature 393:266–268

London ED, Simon SL, Berman SM et al (2004) Mood disturbances and regional cerebral metabolic abnormalities in recently abstinent methamphetamine abusers. Arch Gen Psychiatry 61:73–84

Miyoshi S, Mitsuoka K, Nishimura S, Veltkamp SA (eds) (2011) Radioisotopes in drug research and development: focus on positron emission tomography. Radioisotopes Applications in Bio-Medical Science, InTech, pp 93–114

Montag C, Bey K, Sha P et al (2014) Is it meaningful to distinguish between generalized and specific Internet addiction? Evidence from a cross-cultural study from Germany, Sweden, Taiwan and China. Asia-Pacific Psychiatry; doi: 10.1111/appy.12122 (Epub ahead of print)

Munafo MR, Matheson IJ, Flint J (2007) Association of the DRD2 gene Taq1A polymorphism and alcoholism: a meta-analysis of case-control studies and evidence of publication bias. Mol Psychiatry 12:454–461

Noble EP (2000) Addiction and its reward process through polymorphisms of the D2 dopamine receptor gene: a review. European Psychiatry 15:79–89

Oh K, Kim Y, Jang K (2005) Operation of the internet game addiction center and the eEementary study for development of the internet game addiction scale. Korea Agency Digital Opportunity and Promotion, Seoul

Pallanti S, Haznedar MM, Hollander E et al (2010) Basal Ganglia activity in pathological gambling: a fluorodeoxyglucose-positron emission tomography study. Neuropsychobiology 62:132–138

Park HS, Kim SH, Bang SA et al (2010) Altered regional cerebral glucose metabolism in internet game overusers: a 18F-fluorodeoxyglucose positron emission tomography study. CNS Spectr 15:159–166

Patton JH, Stanford MS, Barratt ES (1995) Factor structure of the Barratt impulsiveness scale. J Clin Psychol 51:768–774

Potenza MN (2006) Should addictive disorders include non-substance-related conditions? Addiction 101:142–151

Reuter J, Raedler T, Rose M et al (2005) Pathological gambling is linked to reduced activation of the mesolimbic reward system. Nat Neurosci 8:147–148

Robbins TW, Everitt BJ (1999) Drug addiction: bad habits add up. Nature 398:567–570
Shapira NA, Goldsmith TD, Keck PE et al (2000) Psychiatric features of individuals with problematic internet use. J Affect Disord 57:267–272
Soloff PH, Meltzer CC, Becker C et al (2003) Impulsivity and prefrontal hypometabolism in borderline personality disorder. Psychiatry Res 123:153–163
Thanos PK, Wang GJ, Volkow ND (2008) Positron emission tomography as a tool for studying alcohol abuse. Alcohol Res Health 31:233–237
Volkow ND, Chang L, Wang GJ et al (2001) Low level of brain dopamine D2 receptors in methamphetamine abusers: association with metabolism in the orbitofrontal cortex. Am J Psychiatry 158:2015–2021
Volkow ND, Fowler JS, Wang GJ et al (1993) Decreased dopamine D2 receptor availability is associated with reduced frontal metabolism in cocaine abusers. Synapse 14:169–177
Volkow ND, Fowler JS (2000) Addiction, a disease of compulsion and drive: involvement of the orbitofrontal cortex. Cereb Cortex 10:318–325
Volkow ND, Fowler JS, Wang GJ, Goldstein RZ (2002) Role of dopamine, the frontal cortex and memory circuits in drug addiction: insight from imaging studies. Neurobiol Learn Mem 78:610–624
Volkow ND, Fowler JS, Wang GJ et al (2009) Imaging dopamine's role in drug abuse and addiction. Neuropharmacology 56(Suppl 1):3–8
Volkow ND, Fowler JS, Wolf AP et al (1990) Effects of chronic cocaine abuse on postsynaptic dopamine receptors. Am J Psychiatry 147:719–724
Volkow ND, Wang GJ, Fowler JS (1996) Decreases in dopamine receptors but not in dopamine transporters in alcoholics. Alcohol Clin Exp Res 20:1594–1598
Volkow ND, Wang GJ, Telang F et al (2007) Profound decreases in dopamine release in striatum in detoxified alcoholics: possible orbitofrontal involvement. J Neurosci 27:12700–12706
Volkow ND, Wang GJ, Telang F et al (2008) Low dopamine striatal D2 receptors are associated with prefrontal metabolism in obese subjects: possible contributing factors. Neuroimage 42:1537–1543
Wang GJ, Volkow ND, Logan J et al (2001) Brain dopamine and obesity. Lancet 357:354–357
Wang GJ, Volkow ND, Fowler JS et al (1997) Dopamine D2 receptor availability in opiate-dependent subjects before and after naloxone-precipitated withdrawal. Neuropsychopharmacology 16:174–182
Yen JY, Ko CH, Yen CF et al (2007) The comorbid psychiatric symptoms of Internet addiction: attention deficit and hyperactivity disorder (ADHD), depression, social phobia, and hostility. J Adolesc Health 41:93–98
Young K (1998) Caught in the net: how to recognize the signs of internet addiction–and a winning strategy for recovery. Wiley, New York
Young K (2010) Internet addiction over the decade: a personal look back. World Psychiatry 9:91

Chapter 5
Functional Brain Changes in Response to Treatment of Internet Gaming Disorder

Doug Hyun Han, Sun Mi Kim and Perry F. Renshaw

Abstract This chapter consists of three parts. First, we suggest the possible therapeutic mechanisms of dopaminergic and serotonergic medications for Internet gaming disorder (IGD) through an overview of the results of pharmacological trials. Second, we review the functional brain changes, especially in terms of cortico-striatal circuitry, in response to pharmacological treatments in individuals with IGD. Finally, we discuss the functional brain changes observed in response to non-pharmacological interventions, including family therapy and abstinence from online gaming. Although the studies discussed in this chapter suggest that a partial recovery of dysfunctional brain activity may be possible, the existence of a causal relationship between brain functional abnormalities and IGD is still open to debate. Functional neuroimaging studies with novel and efficient designs are needed for the development of more effective treatments for IGD as well as for increasing the understanding of the pathophysiology of IGD.

5.1 Introduction

In a meta-analysis of 16 studies with appropriate effect sizes, Winkler et al. (2013) concluded that psychological and pharmacological interventions were effective for improving clinical symptoms, time spent online, depressive mood, and anxiety in individuals with internet gaming disorder (IGD) and internet addiction (IA). Here, IA refers to the generalized form of internet addiction, which is problematic internet use, including a broad range of online activities, otherwise, IGD is a specific form of internet addiction. Specific forms of internet addiction refer to

D.H. Han (✉) · S.M. Kim
Department of Psychiatry, College of Medicine, Chung-Ang University, Seoul, Republic of Korea
e-mail: hduk@yahoo.com

P.F. Renshaw
Brain Institute, University of Utah, Salt Lake City, UT, USA

C. Montag and M. Reuter (eds.), *Internet Addiction*, Studies in Neuroscience, Psychology and Behavioral Economics, DOI 10.1007/978-3-319-07242-5_5

the problematic internet use of distinct activities such as online video gaming, online sex, online gambling, online shopping, or online social-networking. The importance of distinguishing between a generalized form—and specific forms—of internet addiction has previously been proposed by Davis (2001) and has received empirical support from a cross-cultural study by Montag et al. (2014). Montag et al. (2014) compared generalized and specific internet addiction and showed that the two were discrete constructs. Thus far, only a limited number of studies have examined the effectiveness of pharmacological treatments for IGD and IA, and even fewer studies have assessed the functional brain changes that occur in response to pharmacological or non-pharmacological interventions. The pharmacological trials have been designed to target impulsivity by regulating synaptic levels of dopamine or serotonin: escitalopram (Dell'Osso et al. 2008), methylphenidate (Han et al. 2009), and bupropion (Han et al. 2010a; Han and Renshaw 2012). As mentioned in the previous chapter regarding functional neuroimaging studies of IGD, biological theories of IGD have focused on possible alterations in cortico-striatal circuitry. Some studies have assessed functional brain changes in response to pharmacological (Han et al. 2010a) or non-pharmacological intervention (Han et al. 2012a; Kim et al. 2012) in individuals with IGD.

In this chapter, we will initially discuss the possible therapeutic mechanism of dopaminergic and serotonergic medications for IGD through an overview of the results of pharmacological trials. Then, we will discuss the functional brain changes, especially in terms of cortico-striatal circuitry, in response to pharmacological treatments in individuals with IGD. Later in this chapter we will discuss the functional brain changes observed in response to non-pharmacological interventions, including family therapy (Han et al. 2012a) and abstinence from online gaming (Kim et al. 2012).

5.2 Dopamine and Serotonin in Internet Gaming Disorder

5.2.1 Dopamine in Internet Gaming Disorder

Pleasure can be triggered by dopamine release in response to external stimuli including alcohol, drugs, gambling, food, sex and risk taking behaviors (Comings and Blum 2000). Urge-driven behaviors in pathologic gamblers are thought to be mediated by dopamine neurons in the mesolimbic pathway (Bechara 2003). Alterations in dopamine release have also been associated with video game play (Koepp and Silver 1998; Schultz et al. 1993). Based on reports that video game play released dopamine in striatal areas (Bechara 2003; Koepp and Silver 1998; Schultz et al. 1993), online game play is thought to induce neuronal release of dopamine in striatal areas, resulting in reduced craving and negative emotions. As a trial for displacement of dopamine release induced by game playing, pharmacological interventions using bupropion (Han and Renshaw 2012) and methylphenidate (Han et al. 2009) were conducted and changes in the severity of clinical symptoms were assessed.

5.2.1.1 Pharmacological Intervention Using Bupropion (Han and Renshaw 2012)

Han et al. (2012b) conducted a 12-week, open label bupropion trial for individuals with IGD comorbid with major depressive disorder. In this study, IGD was defined as excessive time for game play (more than 4 h per day or 30 h per week), scores greater than 50 on the Young Internet Addiction Scale (YIAS) (Young 1996), and maladaptive behaviors and distress resulting from uncontrollable online game play. Eleven participants who met inclusion criteria for IGD and eight healthy comparison participants who had experience of playing StarCraft for less than 3 days per week and 1 h per day were recruited. During a 12-week prospective trial, including 8 weeks of active treatment and 4 weeks of post treatment follow-up, bupropion was shown to improve depressive mood as well as to reduce the severity of problematic online game play. During the active treatment period, mean YIAS scores were significantly reduced from 71.2 to 45.2 and mean Beck Depression Inventory (BDI) scores were also significantly decreased from 27.6 to 17.7. In addition, mean Clinical Global Impressions-Severity scale scores (CGI-S) (Guy 1976) improved significantly from 3.7 to 1.7 (lower scores indicating an improvement). During the 4-week post-treatment follow-up period, there were no significant changes in mean YIAS scores (from 45.2 to 42.4) while mean BDI scores were increased from 17.7 to 20.5. There was a small and statistically insignificant increase in mean CGI-S scores from 1.7 to 2.0.

Dysfunction in dopaminergic neurotransmission may be associated with dysregulation of reward seeking behavior and subsequent substance use disorders or impulse control disorders (Bowirrat and Oscar-Berman 2005). Increased dopamine release within the brain reward system can reduce negative emotions due to withdrawal symptom and reduce craving for addictive substances or behaviors. Bupropion treatment is effective, through dual norepinephrine and dopamine reuptake inhibition (Cooper et al. 1980, 1994), in reducing craving and relapse rates in individuals with nicotine dependence (Durcan et al. 2002; Hays et al. 2009), cocaine dependence (Margolin et al. 1995), as well as pathologic gambling (Dannon et al. 2005). Han et al. (2012) reported that the bupropion improved clinical symptoms in individuals with IGD and major depressive disorder. Increased extracellular concentrations of dopamine as a result of bupropion treatment may replace dopamine release originally induced by online game play in individuals with IGD. Subsequently, dopamine replacement results in reduced negative emotion and craving for gaming during abstinence.

5.2.1.2 Pharmacological Intervention Using Methylphenidate (Han et al. 2009)

Additional evidence for the effectiveness of dopamine replacement was provided by a study of methylphenidate treatment of children with attention deficit hyperactivity disorder (ADHD) who had been playing online video games (Han et al. 2009). Sixty-two drug-naive children diagnosed with ADHD and excessive online video game play participated in this study. Eight weeks of methylphenidate treatment reduced the severity of attentional dysfunction, assessed using DuPaul's ADHD Rating Scale, Parent and Teacher Version (DuPaul 1991) from a mean of 43.0–32.6. In addition, mean YIAS scores were also significantly reduced from 54.0 to 41.2. Changes in YIAS scores during treatment were positively correlated with changes in DuPaul's ADHD Rating Scale scores.

Reduced dopaminergic neurotransmission in the cortico-striatal pathway, which may reflect prefrontal dysfunction in individuals with ADHD, has been consistently reported (Volkow et al. 2007). In addition, it has been suggested that individuals with ADHD use nicotine and alcohol as self-medication in order to induce dopaminergic activity (Levin et al. 1996; Mihailescu and Drucker-Colin 2000). The blockade of dopamine transporters (DAT) by methylphenidate activates dopaminergic neurotransmission and enhances task-specific signaling (Volkow et al. 2007). Additionally, oral methylphenidate activates dopaminergic neurotransmission, as proven by a decrease in dopamine D2 receptor availability in the striatum (Volkow et al. 2001). Prevalence of the TaqIA minor (A1) allele of the D2 dopamine receptor in individuals with IGD was investigated (Han et al. 2007). In this study, prevalence of A1+(A1A1 and A1A2 genotypes) allelic carriers was significantly higher in the IGD group, which was associated with higher reward dependency. A1+allelic carriers are known to have 30–40 % fewer mesolimbic D2 receptors (Jonsson et al. 1999; Pohjalainen et al. 1998). Han et al. (2009) have suggested that children with ADHD might use online game play as a form of self-medication in order to improve attention and reduce distractibility by activating cortico-striatal circuitry and inducing dopaminergic neurotransmission.

5.2.2 Serotonin in Internet Gaming Disorder

IGD is characterized by pathological repetitive thoughts and behaviors, which are also observed in pathological gambling as well as in obsessive compulsive disorder. Pathological repetitive behaviors and ruminative thinking are associated with increased impulsivity (Petry 2001). Shapira et al. (2000) evaluated the psychiatric features of 20 individuals with IA and found that all of the participants met criteria for an impulse control disorder—not otherwise specified (ICD-NOS) according to the fourth edition of the Diagnostic and Statistical Manual of Mental Disorders (DSM-IV) (American Psychiatric Association 1994). Similarly, in a clinical setting, individuals with IGD and IA often receive the diagnosis of ICD-NOS because there is no specific diagnosis for these disorders in the DSM-IV, although IGD is identified in Sect. 5.3 of DSM-5 (American Psychiatric Association 2013) as a condition requiring more research in order to be included in Sect. 5.2 as one of the formal disorders. IGD and IA are highly comorbid with depression and social phobia (Ko et al. 2008; Yen et al. 2007). Serotonin is thought to play a major role in depressive mood and anxiety as well as in impulsivity (Lesch and Merschdorf 2000). Selective serotonin reuptake inhibitors (SSRIs) have been widely used to treat both obsessive compulsive disorder and major depressive disorder (Dell'Osso et al. 2005; Greist et al. 1995). In this context, SSRIs may have therapeutic potential for treating IGD and IA.

5.2.2.1 Pharmacological Intervention Using Escitalopram (Dell'Osso et al. 2008)

Dell'Osso et al. (2008) evaluated the effectiveness of escitalopram for the treatment of IA in a two-phase trial: Nineteen subjects were enrolled at a 10-week open-label phase. At the end of 10th week, individuals who completed first phase (seventeen patients) were randomly assigned to the escitalopram group or placebo group for 9 weeks. For this trial, the authors recruited participants with uncontrollable, distressing, and excessive Internet use, which caused social, relational, occupational, and financial problems. At the end of 10-weeks of open-label treatment with escitalopram, a significant decrease in mean time spent online, from 36.8 to 16.5 h per week, was noted. In addition, 64.7 % of participants with IA showed positive therapeutic responses defined as a rating of "*much improved*" or "*very much improved*" on the Clinical Global Impressions-Improvement scale (CGI-I) (Guy 1976). At the end of a 9-week discontinuation phase, both groups taking escitalopram and placebo maintained the effects achieved during the open-label phase without any significant differences. The authors suggested that 9 weeks might be too short for further improvement to be shown in the escitalopram group or for the effect during the open-label phase to be disappeared in the placebo group. These favorable results suggest a role for serotonin dysregulation in IGD, as well as the utility of SSRIs as potential therapeutic agents for IGD in terms of improving impulsivity as well as depression. Further studies using neurobiological assessments are needed to determine the influence of SSRI treatment on functional brain changes associated with IGD.

5.3 Functional Brain Changes in Response to Pharmacological Interventions

As noted in a previous chapter, research on IGD has consistently utilized fMRI paradigms. It has been demonstrated that gaming cues activate the dorsolateral prefrontal cortex (DLPFC), parahippocampus, anterior cingulate, orbitofrontal cortex (OFC), nucleus accumbens, and dorsal striatum in individuals with IGD and IA (Han et al. 2010b, 2011; Ko et al. 2008; Lorenz et al. 2013; Sun et al. 2012). Participant craving for online games has been shown to positively correlate with activation of the brain reward system, including the prefrontal cortex, striatum and amygdala (Han et al. 2010b, 2011; Ko et al. 2008; Sun et al. 2012). The results from these studies may support the hypothesis that online game play induces dopamine release in the brain reward circuitry. In another study conducted by Montag et al. (2012), increased activation of DLPFC and temporal regions of the brain in gamers was demonstrated during an exposure to stills from the game in excessive first person-shooter-video-players. In this case, increased response of DLPFC could reflect the activation of action scripts associated with the game. With the assumption that online game play induces dopamine release in the brain reward circuitry and dopaminergic agents reduce negative emotions and craving for gaming through replacement of dopamine, Han et al. (2010a) have assessed changes in brain activity in response to bupropion treatment in IGD.

5.3.1 Functional Brain Changes in Response to Bupropion Treatment (Han et al. 2010a)

Han et al. (2010a) investigated whether bupropion sustained release (SR) treatment would change the intensity of craving for gaming and gaming cue-induced brain activity in individuals with IGD. The inclusion criteria for the IGD group were: spending a lot of time gaming (i.e. more than 4 h per day or 30 h per week), a score greater than 50 on the YIAS, and maladaptive behaviors and distress due to extensive online video game play. At baseline assessment, individuals with IGD demonstrated greater brain activity when exposed to visual gaming cues in the left occipital lobe cuneus, left DLPFC, and left parahippocampal gyrus, compared to healthy matched comparison subjects. After 6-weeks of bupropion SR treatment, the level of craving for online gaming, the amount of time spent gaming, and the activity of the DLPFC induced by gaming cues were reduced in individuals with IGD. Additionally, the gaming cue-induced activity within the DLPFC was shown to positively correlate with the level of craving for online gaming.

Bupropion SR may change the intensity of craving and brain activity in individuals with IGD in similar ways to those observed in individuals with substance or behavioral addiction (Dannon et al. 2005; Durcan et al. 2002; Hays et al. 2009; Margolin et al. 1995). Decreased gaming cue-induced activity in the DLPFC following a 6 week period of bupropion SR treatment could be explained by the mechanism of action of bupropion, i.e., blocking the reuptake of norepinephrine and dopamine. This mechanism of action of bupropion increases extracellular concentrations of these catecholamines in the frontal cortex, hypothalamus, and nucleus accumbens of rats (Li et al. 2002; Nomikos et al. 1992). In turn, this may result in the reduction of cue-induced surges of norepinephrine and dopamine release within the DLPFC (Dazzi et al. 2001a, b, 2002), an important region in cue-elicited craving (Crockford et al. 2005; Maas et al. 1998). Decreases in cue-induced surges of norepinephrine and dopamine release may attenuate the positive reinforcing properties of gaming, and consequently reduce craving for game playing. In this study, gaming cue-induced activity in the striatum was not significantly changed after 6-week bupropion SR treatment, although a high abundance of dopamine in this area (Ciliax et al. 1999). This could be explained by previous study results indicating that sustained administration of bupropion exerted its therapeutic influence by increasing synaptic availability of norepinephrine rather than dopamine (Dong and Blier 2001; El Mansari et al. 2008). The prefrontal cortex is an important area in which noradrenergic system exerts its effects by promoting cognitive abilities, including response inhibition, behavioral flexibility, and attention (Cole and Robbins 1992; Lapiz and Morilak 2006).

In the same context, bupropion treatment has been reported to reduce impulsivity and increase attention in children with ADHD, who are thought to have functional deficits in the DLPFC (Barrickman et al. 1995; Kuperman et al. 2001; Wilens et al. 2001). The therapeutic effects of bupropion in ADHD, i.e. reducing impulsivity and improving attention, may be consistent with those of methylphenidate treatment on internet video game play in children with ADHD (Han et al. 2009), which we discussed earlier in this chapter, in terms of increasing dopaminergic neurotransmission within the cortico-striatal pathway.

Further mechanisms for bupropion's effectiveness in IGD may rely on its role as a norepinephrine reuptake inhibitor. Bupropion is known to be effective for alleviation of withdrawal symptoms in cocaine dependence, including fatigue, psychomotor retardation, and hypersomnia, by increasing noradrenergic activity (Ascher et al. 1995; Cooper et al. 1980, 1994). Additionally, bupropion is reported to improve negative affect during smoking cessation (Lerman et al. 2002). This mechanism may increase the effectiveness of bupropion treatment in IGD, in terms of alleviating negative affect during abstinence, through norepinephrine reuptake inhibitor inhibition, which results in reduced levels of craving for gaming.

5.4 Functional Brain Changes in Response to Non-pharmacological Interventions

A few studies have assessed functional brain changes in response to non-pharmacological interventions, although various types of intervention have been reported to be effective for improving the clinical symptoms of IGD and IA, including cognitive behavior therapy, family therapy, as well as parent and teacher education (Winkler et al. 2013). Based on previous results reporting an association between dysfunctional family structure and adolescent substance abuse (Fröjd et al. 2007; Roustit et al. 2007), and an association between dopaminergic reward circuitry and recognition of maternal and romantic love (Bartels and Zeki 2004; Fisher et al. 2005), Han et al. (2012) conducted an fMRI study regarding the effect of family therapy on IGD. In another study, Kim et al. (2012) evaluated whether a short abstinence period from online game play would change brain activity induced by working memory tasks in IGD, based on previous studies reporting deficits in working memory in adolescents with substance dependence (Tapert et al. 2002; Tapert et al. 2004).

5.4.1 Functional Brain Changes in Response to Family Therapy (Han et al. 2012a)

Han et al. (2012a) assessed the effects of a 3-week family therapy intervention on the severity of IGD as well as on the patterns of brain activity when exposed to visual cues depicting scenes of parental affection and cues depicting online game scenes in adolescents with IGD. For this study, 15 adolescents with IGD and their families who reported moderate to severe family dysfunction were recruited. Inclusion criteria for participants with IGD were excessive time spent gaming (more than 4 h per day or 30 h per week), scores greater than 50 on the YIAS, and maladaptive behaviors and distress due to uncontrollable online game play. Members from these fifteen families were requested to complete homework assignments focused on cultivating family cohesion for more than 1 h day, 4 days a week, over a 3 week period. At baseline, adolescents with IGD showed decreased activity in the caudate, middle temporal gyrus, and occipital lobe in response to cues depicting scenes of parental affection and increased activity in the middle frontal and inferior parietal cortices in response to gaming cues, compared to healthy adolescents with relatively functional family structures. At follow up assessment, adolescents with IGD reported improvement in the perceived level of family cohesion. The changes in the perceived level of family cohesion were positively correlated with an increase in the activity of the caudate nucleus in response to cues depicting scenes of parental affection, and were negatively associated with changes in time spent gaming. Conversely, activity within the DLPFC in response to gaming cues was reduced following 3-weeks of family therapy.

Craving for romantic and parental love and craving for gaming in individuals with IGD might, in part, share the same biological underpinnings. Brain reward pathways, including the ventral tegmental area and the caudate nucleus, are thought to mediate the response to stimuli depicting romantic and parental love (Bartels and Zeki 2004; Fisher et al. 2005). Aron et al. (2005) reported increased activity in the right caudate in response to images of lovers in the early stage of romantic passion. In addition, Vrticka et al. (2008) reported that secure adult attachment style was associated with an increased activation of striatum and the ventral tegmental area in response to stimuli regarding positive social relationships, such as a smiling face. The characteristics of love, such as particular attention to someone, increased drive, excited mood, as well as desire to have a close relationship with a partner, are associated with dopamine release in the brain reward system (Aron et al. 2005; Bartels and Zeki 2004). Additionally, it has been reported that college students with poor parental care during early life showed increased dopamine release in ventral striatum in response to stress-inducing tasks, compared to students who experienced good parental care (Pruessner et al. 2004). An excessive dopamine release in response to substances or stressful events is thought to contribute vulnerability to substance dependence (Marinelli and Piazza 2002; Piazza et al. 1998). In this context, Han et al. (2012) suggested that adolescents with problematic IGD in dysfunctional families tended to have reduced salience for parent-child relationships. In addition, the authors suggested that adolescents with low levels of perceived family cohesion might play online games in order to compensate for striatal dopamine deficits related to poor parental care during early life. Increases in the perceived level of family cohesion and parental love during family therapy may facilitate dopamine release in brain reward circuits and, in turn, this may help reduce the level of craving for gaming.

5.4.2 Functional Brain Changes in Response to Abstinence from Online Game Play (Kim et al. 2012)

Kim et al. (2012) evaluated whether 4 weeks of abstinence from online game would change brain activity in response to simple and complex working memory tasks in adolescents with IGD. Adolescents who reported extensive time spent on gaming (more than 4 h per day or 30 h per week), scores greater than 50 on the YIAS, and distress due to excessive online game play, were recruited in this study. At baseline, adolescents with IGD demonstrated increased activity in response to working memory tasks within the right middle occipital gyrus, left cerebellar posterior lobe, left premotor cortex and left middle temporal gyrus, relative to healthy comparison subjects. After 4-weeks of abstinence from online gaming, adolescents with IGD showed increased activity in the right DLPFC, and left occipital fusiform gyrus when exposed to working memory tasks. In addition, reductions in the severity of IGD were correlated with increases in functional changes in the right DLPFC in response to complex working memory tasks.

The DLPFC is known to play a central role for working memory, including selective attention, inhibition of perseverative errors, and impulse control (Kane and Engle 2002; Leiserson and Pihl 2007). The major involvement of the DLPFC in working memory performance and its activation when human subjects carry out working memory tasks have been well established by neuroimaging studies (Goldman-Rakic 1995). In addition, working memory deficits and correlated dysfunction in the DLPFC have been shown in various neuropathological conditions, including schizophrenia, Parkinson's disease, as well as age-related cognitive decline (Goldman-Rakic 1995). It has been reported that DLPFC activity in response to working memory tasks increases during short-term abstinence from cannabis dependence (Pope et al. 2001; Yurgelun-Todd et al. 1998). Kim et al. (2012) suggested that the dysfunction of working memory in individuals with IGD might be similar to that observed in individuals with substance dependence. Based on the result of those studies, excessive game play might impede the appropriate function of the DLPFC on working memory tasks. However, there has been a large strain of research that shows gaming can have positive effects of working memory. Several studies have reported that action video games seem to improve visuospatial working memory capacity, selective attention, and task-switching ability in healthy individuals (Bavelier et al. 2012; Green and Bavelier 2003; Green et al. 2012). In addition, as we discussed earlier, Han et al. (2009) suggested that children with ADHD might use online game for enhancing task-specific signaling by compensating for striatal dopamine deficits. It seems that the beneficial or harmful effects of gaming on working memory are determined by both the characteristics of a gamer, including casual use and problematic use, and the genre of a game. In addition, methodological limitations, such as cross-sectional comparisons of gamers and non-gamers, make it impossible to distinguish baseline differences in cognitive skills from the consequences of gaming. Further studies are needed to understand effects of gaming on working memory and related cognitive abilities.

5.5 Future Studies of Functional Brain Changes in Response to Treatment

Although the studies discussed in this chapter imply that a partial recovery of dysfunctional brain activity may be possible, the existence of a causal relationship between brain functional abnormalities and IGD is still open to debate. Distinguishing underlying vulnerability from the consequences of excessive Internet and online game use is important in order to better understand the pathophysiology of—and to develop improved treatment strategies for—IGD. As part of an effort to investigate a possible causal relationship between IGD and brain functional abnormalities, Han et al. (2012a) compared the regional brain volumes of individuals with IGD to those of professional gamers. Although professional gamers play online games extensively, they are able to regulate their game use and do not have functional impairment in their daily routine. In this study, professional gamers were characterized by increased gray matter volume in the left cingulate gyrus in comparison with individuals with IGD, as well with healthy comparison subjects. The gray matter volume of the left cingulate gyrus was negatively correlated with levels of impulsivity and perseverative errors in professional gamers. Larger gray matter volumes in the left anterior cingulate could be associated with superior attentional control and decision-making in professional gamers, which might predispose them to successful game play. Decreased gray matter volume and reduced baseline regional cerebral blood flow in the anterior cingulate were reported in individuals with substance dependence (Childress et al. 1999; Franklin et al. 2002). During an exposure to substance-related stimuli, substance users demonstrated increased cerebral blood flow in the anterior cingulate (Brody et al. 2007; Childress et al. 1999). The anterior cingulate cortex is known to play an important role in monitoring and resolving conflict. Individuals with substance addiction are prone to be confronted with the conflict between seeking immediate reward (e.g. taking the drug) and avoiding long-term negative consequences (Bush et al. 2000; Pochon et al. 2008).

Future studies comparing functional brain changes in individuals with IGD to those of professional gamers may also be helpful for distinguishing underlying vulnerability from the consequences of IGD in terms of neurobiology. Imaging genetics, which refer to combining molecular genetics with neuroimaging, would help to further disentangle the vulnerability to—and consequences—of gaming. This technique makes it possible to deduce how genetic variants impact certain brain areas, both structurally and functionally, from the obtained data (de Geus et al. 2008). A good example is a study by Montag et al. (2010), comprising an imaging genetics study in the field of addiction research. The interaction effect of functional variants on brain derived neurotrophic factor (BDNF) and dopamine receptor D2 (DRD2) genes on gray matter volume of the anterior cingulate was demonstrated. These efforts would also facilitate the development of effective therapies focused on reducing or eliminating vulnerability factors for IGD.

5.6 Conclusion

According to the results of fMRI studies conducted with pharmacological and non-pharmacological therapeutic interventions in individuals with IGD, increases in activity in the DLPFC and in dopaminergic neurotransmission within the cortico-striatal pathway while not playing seem to be significantly related to positive therapeutic outcomes. However, because of the limited number of studies to date, it would be premature to draw any firm conclusions regarding functional brain changes in response to treatment of IGD and IA. Further pharmacological trials with placebo-controlled, double blind designs using neurobiological assessments are needed in order to investigate the influence of medication on functional brain change in IGD and IA. Furthermore, non-pharmacological trials with functional neuroimaging methods using relevant stimulation paradigms would be helpful for evaluating the specific effects and strength of different treatment programs. Functional neuroimaging studies with novel and efficient designs are needed for the development of more effective treatments for IGD and IA as well as for increasing understanding of the pathophysiology of IGD and IA. For specific forms of internet addiction other than IGD, such as problematic online sexual behavior, online shopping, and online gambling, therapeutic interventions for each disorder in real life (offline) could be considered first. For example, pharmacotherapy using serotonin reuptake inhibitors, opioid antagonists, and mood stabilizers, as well as cognitive behavior therapy, are known to be effective in the treatment of pathological gambling (Gooding and Tarrier 2009; Sood et al. 2003). Cognitive behavioral and hormonal therapies for the treatment of sexually compulsive behavior (Codispoti 2008) and group cognitive behavioral therapy for the treatment of compulsive buying (Müller et al. 2013), are also thought to be promising treatment strategies. The pharmacological and behavioral therapeutic interventions for these disorders in real life could be also effective for directly-related, specific internet addictions.

References

American Psychiatric Association (1994) Diagnostic and statistical manual of mental disorders, 4th edn. American Psychiatric Association, Washington D.C.

American Psychiatric Association (2013) Diagnostic and statistical manual of mental disorders, 5th edn. American Psychiatric Association, Washington D.C.

Aron A, Fisher H, Mashek DJ et al (2005) Reward, motivation, and emotion systems associated with early-stage intense romantic love. J Neurophysiol 94:327–337

Ascher JA, Cole JO, Colin JN, Feighner JP et al (1995) Bupropion: a review of its mechanism of antidepressant activity. J Clin Psychiatry 56:395–401

Barrickman LL, Perry PJ, Allen AJ et al (1995) Bupropion versus methylphenidate in the treatment of attention-deficit hyperactivity disorder. J Am Acad Child Adolesc Psychiatry 34:649–657

Bartels A, Zeki S (2004) The neural correlates of maternal and romantic love. Neuroimage 21:1155–1166

Bavelier D, Achtman RL, Mani M, Focker J (2012) Neural bases of selective attention in action video game players. Vision Res 61:132–143

Bechara A (2003) Risky business: emotion, decision-making, and addiction. J Gambl Stud 19:23–51

Bowirrat A, Oscar-Berman M (2005) Relationship between dopaminergic neurotransmission, alcoholism, and reward deficiency syndrome. Neuropsychiatric Genet 132:529–537

Brody AL, Mandelkern MA, Olmstead RE et al (2007) Neural substrates of resisting craving during cigarette cue exposure. Biol Psychiatry 62:642–651

Bush G, Luu P, Posner MI (2000) Cognitive and emotional influences in anterior cingulate cortex. Trends Cogn Sci 4:215–222

Childress AR, Mozley PD, McElgin W et al (1999) Limbic activation during cue-induced cocaine craving. Am J Psychiatry 156:11–18

Ciliax BJ, Drash GW, Staley JK et al (1999) Immunocytochemical localization of the dopamine transporter in human brain. J Comp Neurol 409:38–56

Codispoti VL (2008) Pharmacology of sexually compulsive behavior. Psychiatr Clin North America 31:671–679

Cole BJ, Robbins TW (1992) Forebrain norepinephrine: role in controlled information processing in the rat. Neuropsychopharmacology 7:129–142

Comings DE, Blum K (2000) Reward deficiency syndrome: genetic aspects of behavioral disorders. Prog Brain Res 126:325–341

Cooper BR, Hester TJ, Maxwell RA (1980) Behavioral and biochemical effects of the antidepressant bupropion (Wellbutrin): evidence for selective blockade of dopamine uptake in vivo. J Pharmacol Exp Ther 215:127–134

Cooper BR, Wang CM, Cox RF et al (1994) Evidence that the acute behavioral and electrophysiological effects of bupropion (Wellbutrin) are mediated by a noradrenergic mechanism. Neuropsychopharmacology 11:133–141

Crockford DN, Goodyear B, Edwards J (2005) Cue-induced brain activity in pathological gamblers. Biol Psychiatry 58:787–795

Dannon PN, Lowengrub K, Musin E (2005) Sustained-release bupropion versus naltrexone in the treatment of pathological gambling: a preliminary blind-rater study. J Clin Psychopharmacol 25:593–596

Davis RA (2001) A cognitive-behavioral model of pathological Internet use. Comput Hum Behav 17:187–195

Dazzi L, VignoneV Seu E (2002) Inhibition by venlafaxine of the increase in norepinephrine output in rat prefrontal cortex elicited by acute stress or by the anxiogenic drug FG 7142. J Psychopharmacol 16:125–131

Dazzi L, Serra M, Spiga F (2001a) Prevention of the stress-induced increase in frontal cortical dopamine efflux of freely moving rats by long-term treatment with antidepressant drugs. Eur Neuropsychopharmacol 11:343–349

Dazzi L, Spiga F, Pira L (2001b) Inhibition of stress- or anxiogenic-drug-induced increases in dopamine release in the rat prefrontal cortex by long-term treatment with antidepressant drugs. J Neurochem 76:1212–1220

De Geus E, Goldberg T, Boomsma DI, Posthuma D (2008) Imaging the genetics of brain structure and function. Biol Psychol 79:1–8

Dell'Osso B, Allen A, Hollander E (2005) Fluvoxamine: a selective serotonin re-uptake inhibitor for the treatment of obsessive-compulsive disorder. Expert Opin Pharmacother 6:2727–2740

Dell'Osso B, Hadley S, Allen A et al (2008) Escitalopram in the treatment of impulsive-compulsive internet usage disorder: an open-label trial followed by a double-blind discontinuation phase. J Clin Psychiatry 69:452–456

Dong J, Blier P (2001) Modification of norepinephrine and serotonin, but not dopamine, neuron firing by sustained bupropion treatment. Psychopharmacology 155:52–57

DuPaul GJ (1991) Parent and teacher ratings of ADHD symptoms: psychometric properties in a community-based sample. J Clin Child Adolesc Psychol 20:245–253

Durcan MJ, Deener G, White J et al (2002) The effect of bupropion sustained-release on cigarette craving after smoking cessation. Clin Ther 24:540–551
El Mansari M, Ghanbari R, Janssen S (2008) Sustained administration of bupropion alters the neuronal activity of serotonin, norepinephrine but not dopamine neurons in the rat brain. Neuropharmacology 55:1191–1198
Fisher H, Aron A, Brown LL (2005) Romantic love: an fMRI study of a neural mechanism for mate choice. J Comp Neurol 493:58–62
Franklin TR, Acton PD, Maldjian JA et al (2002) Decreased gray matter concentration in the insular, orbitofrontal, cingulate, and temporal cortices of cocaine patients. Biol Psychiatry 51:134–142
Fröjd S, Kaltiala-Heino R, Rimpelä M (2007) The association of parental monitoring and family structure with diverse maladjustment outcomes in middle adolescent boys and girls. Nord J Psychiatry 61:296–303
Goldman-Rakic PS (1995) Cellular basis of working memory. Neuron 14:477–485
Gooding P, Tarrier N (2009) A systematic review and meta-analysis of cognitive-behavioural interventions to reduce problem gambling: hedging our bets? Behav Res Ther 47:592–607
Green CS, Bavelier D (2003) Action video game modifies visual selective attention. Nature 423:534–537
Green CS, Sugarman MA, Medford K (2012) The effect of action video game experience on task-switching. Comput Hum Behav 28:984–994
Greist JH, Jefferson JW, Kobak KA (1995) Efficacy and tolerability of serotonin transport inhibitors in obsessive-compulsive disorder. A meta-analysis. Arch Gen Psychiatry 52:53–60
Guy W (1976) Clinical global impression scale. In: M. Rockville (ed.) The ECDEU assessment manual for psychopharmacology-revised. Department of Health and Human Services, Public Health Service, Alcohol Drug Abuse and Mental Health Administration, NIMH Psychopharmacology Research Branch, US, pp 218–222
Han DH, Renshaw PF (2012) Bupropion in the treatment of problematic online game play in patients with major depressive disorder. J Psychopharmacol 26:689–696
Han DH, Lee YS, Yang KC (2007) Dopamine genes and reward dependence in adolescents with excessive internet video game play. J Addict Med 1:133–138
Han DH, Lee YS, Na C (2009) The effect of methylphenidate on Internet video game play in children with attention-deficit/hyperactivity disorder. Compr Psychiatry 50:251–256
Han DH, Hwang JW, Renshaw PF (2010a) Bupropion sustained release treatment decreases craving for video games and cue-induced brain activity in patients with Internet video game addiction. Exp Clin Psychopharmacol 18: 297–304
Han DH, Kim YS, Lee YS (2010b) Changes in cue-induced, prefrontal cortex activity with video-game play. Cyberpsychol Beha Soc Networking 13:655–661
Han DH, Bolo N, Daniels MA (2011) Brain activity and desire for Internet video game play. Compr Psychiatry 52:88–95
Han DH, Kim SM, Lee YS (2012a) The effect of family therapy on the changes in the severity of on-line game play and brain activity in adolescents with on-line game addiction. Psychiatry Res Neuroimaging 202:126–131
Han DH, Lyoo IK, Renshaw PF (2012b) Differential regional gray matter volumes in patients with on-line game addiction and professional gamers. J Psychiatr Res 46:507–515
Hays JT, Hurt RD, Decker PA (2009) A randomized, controlled trial of bupropion sustained-release for preventing tobacco relapse in recovering alcoholics. Nicotine Tob Res 11:859–867
Jonsson EG, Nothen MM, Grunhage F et al (1999) Polymorphisms in the dopamine D2 receptor gene and their relationships to striatal dopamine receptor density of healthy volunteers. Mol Psychiatry 4:290–296
Kane MJ, Engle RW (2002) The role of prefrontal cortex in working-memory capacity, executive attention, and general fluid intelligence: an individual-differences perspective. Psychon Bull Rev 9:637–671
Kim SM, Han DH, Lee YS (2012) Changes in brain activity in response to problem solving during the abstinence from online game play. J Behav Addict 1:41–49

Ko CH, Yen JY, Chen CS et al (2008) Psychiatric comorbidity of internet addiction in college students: an interview study. CNS Spectr 13:147–153

Koepp DM, Silver PA (1998) Nucleocytoplasmic transport and cell proliferation. Biochim Biophys Acta 17:39–47

Kuperman S, Perry PJ, Gaffney GR (2001) Bupropion SR vs. methylphenidate vs. placebo for attention deficit hyperactivity disorder in adults. Ann Clin Psychiatry 13:129–134

Lapiz MD, Morilak DA (2006) Noradrenergic modulation of cognitive function in rat medial prefrontal cortex as measured by attentional set shifting capability. Neuroscience 137:1039–1049

Leiserson V, Pihl RO (2007) Reward-sensitivity, inhibition of reward-seeking, and dorsolateral prefrontal working memory function in problem gamblers not in treatment. J Gambl Stud 23:435–455

Lerman C, Roth D, Kaufmann V (2002) Mediating mechanisms for the impact of bupropion in smoking cessation treatment. Drug Alcohol Depend 67:219–223

Lesch KP, Merschdorf U (2000) Impulsivity, aggression, and serotonin: a molecular psychobiological perspective. Behav Sci Law 18:581–604

Levin ED, Conners CK, Sparrow E et al (1996) Nicotine effects on adults with attention-deficit/hyperactivity disorder. Psychopharmacology 123:55–63

Li SX, Perry KW, Wong DT (2002) Influence of fluoxetine on the ability of bupropion to modulate extracellular dopamine and norepinephrine concentrations in three mesocorticolimbic areas of rats. Neuropharmacology 42:181–190

Lorenz RC, Kruger JK, Neumann B (2013) Cue reactivity and its inhibition in pathological computer game players. Addict Biol 18:134–146

Maas LC, Lukas SE, Kaufman MJ (1998) Functional magnetic resonance imaging of human brain activation during cue-induced cocaine craving. Am J Psychiatry 155:124–126

Margolin A, Kosten TR, Avants SK et al (1995) A multicenter trial of bupropion for cocaine dependence in methadone-maintained patients. Drug Alcohol Depend 40:125–131

Marinelli M, Piazza PV (2002) Interaction between glucocorticoid hormones, stress and psychostimulant drugs. Eur J Neurosci 16:387–394

Mihailescu S, Drucker-Colin R (2000) Nicotine, brain nicotinic receptors, and neuropsychiatric disorders. Arch Med Res 31:131–144

Montag C, Weber B, Jentgens E et al (2010) An epistasis effect of functional variants on the BDNF and DRD2 genes modulates gray matter volume of the anterior cingulate cortex in healthy humans. Neuropsychologia 48:1016–1021

Montag C, Weber B, Trautner P et al (2012) Does excessive play of violent first-person-shooter-video-games dampen brain activity in response to emotional stimuli? Biol Psychol 89:107–111

Montag C, Bey K, Sha P et al. (2014) Is it meaningful to distinguish between generalized and specific Internet addiction? Evidence from a cross-cultural study from Germany, Sweden, Taiwan and China. Asia-Pacific Psychiatry (official journal of the Pacific Rim College of Psychiatrists)

Müller A, Mitchell JE, de Zwaan M (2013) Compulsive buying. Am J Addict Am Acad Psychiatrists Alcohol Addict

Nomikos GG, Damsma G, Wenkstern D, Fibiger HC (1992) Effects of chronic bupropion on interstitial concentrations of dopamine in rat nucleus accumbens and striatum. Neuropsychopharmacology 7:7–14

Petry NM (2001) Substance abuse, pathological gambling, and impulsiveness. Drug Alcohol Depend 63:29–38

Piazza PV, Deroche V, Rouge-Pont F, Le Moal M (1998) Behavioral and biological factors associated with individual vulnerability to psychostimulant abuse. NIDA Res Monogr 169:105–133

Pochon JB, Riis J, Sanfey AG et al (2008) Functional imaging of decision conflict. J Neurosci 28:3468–3473

Pohjalainen T, Rinne JO, Nagren K et al (1998) The A1 allele of the human D2 dopamine receptor gene predicts low D2 receptor availability in healthy volunteers. Mol Psychiatry 3:256–260

Pope HG Jr, Gruber AJ, Hudson JI (2001) Neuropsychological performance in long-term cannabis users. Arch Gen Psychiatry 58:909–915

Pruessner JC, Champagne F, Meaney MJ, Dagher A (2004) Dopamine release in response to a psychological stress in humans and its relationship to early life maternal care: a positron emission tomography study using [11C] raclopride. J Neurosci 24:2825–2831

Roustit C, Chaix B, Chauvin P (2007) Family breakup and adolescents' psychosocial maladjustment: public health implications of family disruptions. Pediatrics 120:e984–e991

Schultz SC, Grady B, Cole F et al (1993) A role for endothelin and nitric oxide in the pressor response to diaspirin crosslinked hemoglobin. J Lab Clin Med 122:301–308

Shapira NA, Goldsmith TD, Keck PE et al (2000) Psychiatric features of individuals with problematic internet use. J Affect Disord 57:267–272

Sood ED, Pallanti S, Hollander E (2003) Diagnosis and treatment of pathologic gambling. Curr Psychiatry Rep 5:9–15

Sun Y, Ying H, Seetohul RM et al (2012) Brain fMRI study of crave induced by cue pictures in online game addicts (male adolescents). Behav Brain Res 233:563–576

Tapert SF, Granholm E, Leedy NG, Brown SA (2002) Substance use and withdrawal: neuropsychological functioning over 8 years in youth. J Int Neuropsychol Soc 8:873–883

Tapert SF, Schweinsburg AD, Barlett VC (2004) Blood oxygen level dependent response and spatial working memory in adolescents with alcohol use disorders. Alcohol Clin Exp Res 28:1577–1586

Volkow ND, Wang G, Fowler JS (2001) Therapeutic doses of oral methylphenidate significantly increase extracellular dopamine in the human brain. J Neurosci 21:1–5

Volkow ND, Wang GJ, Newcorn J et al (2007) Brain dopamine transporter levels in treatment and drug naive adults with ADHD. Neuroimage 34:1182–1190

Vrticka P, Andersson F, Grandjean D et al (2008) Individual attachment style modulates human amygdala and striatum activation during social appraisal. PLoS ONE 3:0002868

Wilens TE, Spencer TJ, Biederman J et al (2001) A controlled clinical trial of bupropion for attention deficit hyperactivity disorder in adults. Am J Psychiatry 158:282–288

Winkler A, Dorsing B, Rief W (2013) Treatment of internet addiction: a meta-analysis. Clin Psychol Rev 33:317–329

Yen JY, Ko CH, Yen CF et al (2007) The comorbid psychiatric symptoms of Internet addiction: attention deficit and hyperactivity disorder (ADHD), depression, social phobia, and hostility. J Adolesc Health 41:93–98

Young KS (1996) Psychology of computer use: XL. addictive use of the internet: a case that breaks the stereotype. Psychol Rep 79:899–902

Yurgelun-Todd D, Gruber S, Hanson R (1998) Residual effects of marijuana use: an fMRI study. Paper presented at the The 60th annual scientific meeting, the college on problems of drug dependence. NIDA Research Monograph 179, National Institute on Drug Abuse, Bethesda, MD

Chapter 6
Molecular Genetics, Personality and Internet Addiction

Christian Montag and Martin Reuter

Abstract Mounting evidence suggests that personality plays an important role for a better understanding of Internet addiction. As twin studies have shown that both Internet addiction and personality are influenced by genetics, we highlight molecular genetics approaches to Internet addiction in this chapter too.

6.1 Background of the Research Question

Internet addiction has become a global phenomenon. While most of the early studies have been conducted in Asian countries, strong research efforts can increasingly be observed in Western countries. Although prevalence rates strongly differ between countries/cultures (Shaw and Black 2008), Internet addicts are a strong matter for debate among scientific circles and practitioners around the globe. In Germany, current estimates derived from a representative sample state a prevalence of about 1 % (Rumpf et al. 2011).

The inclusion of Internet Gaming Disorder in section III of DSM-V as an emerging disorder underlines the importance to further characterize Internet addiction by conducting new research in this area. Clearly, the inclusion of Internet Gaming Disorder in DSM-V represents just a first step towards an acceptance of this rising

M. Reuter
Department of Psychology, University of Bonn, Bonn, Germany

M. Reuter
Laboratory of Neurogenetics, University of Bonn, Bonn, Germany

M. Reuter
Center for Economics and Neuroscience, University of Bonn, Bonn, Germany

C. Montag (✉)
Department of Psychology, University of Ulm, Ulm, Germany
e-mail: christian.montag@uni-ulm.de

C. Montag and M. Reuter (eds.), *Internet Addiction*, Studies in Neuroscience, Psychology and Behavioral Economics, DOI 10.1007/978-3-319-07242-5_6

problem, because Internet Gaming Disorder overlaps only in part with generalized Internet addiction (Montag et al. 2014), which can best be characterized by addiction to the Internet in more broad terms (spending lots of time online in chats, checking e-mails excessively, procrastinating over every day work by spending too much time on the Internet). Moreover, it needs to be mentioned, that the empirical evidence on the nature of Internet addiction, as it stands, is currently multifaceted (Ko et al. 2012). Although researchers do not agree on standard symptoms for an Internet addiction diagnosis, a 2 + 1 rule has been put forward by Tao et al. (2010). Based on empirical data, Tao et al. suggested that both 'withdrawal symptoms when not being online' and 'preoccupation with the Internet' are prerequisite for Internet addiction and must be accompanied by at least one further symptom, such as development of tolerance. Besides a large body of classic psychological studies dealing with questions on how personality (e.g. Ko et al. 2006; Chak and Leung 2004; Cao et al. 2007; Montag et al. 2010, 2011; Sariyska et al. 2014) or sociodemographic factors (in particular gender, age) are linked to Internet addiction (e.g. Hur 2006; Tsai et al. 2009), more and more studies are incorporating neuroscientific tools, including functional and structural magnetic resonance imaging, to shed light on the actual nature of Internet addiction (e.g. Dong et al. 2010; Yuan et al. 2011; Zhou et al. 2011).

The present chapter aims to provide a short review of two important related aspects, which may be considered vulnerability factors for Internet addiction—namely personality traits and molecular genetic factors. As most studies in these two fields used cross-sectional designs, clearly no causality can be derived from the majority of findings. Nevertheless, conceptualization of the term 'personality' and of genetic factors suggests that these research targets represent a cause for—rather than an effect of—Internet addiction. This will be further explained in the next sections, followed by a summary of the empirical evidence on personality/genetics and Internet addiction.

6.2 What Is Personality?

Definitions for personality are numerous and these definitions differ across personality theories. For an introduction to several important (in particular biological) personality theories, two of our own review studies are particularly recommended (see Montag et al. 2012, 2013). Despite the numerous approaches to personality, the term needs to be briefly defined to enable a better understanding of the present chapter.

In our opinion, the most important common denominator in defining personality across theories can be found in the term *trait*. A personality *trait* represents an enduring disposition of a person to behave with certain behavioral, cognitive and emotional tendencies to heterogeneous demands across a large number of situations. For example, individual differences in personality traits can manifest in being overtly anxious or socially outgoing when dealing with diverse problems/situations in everyday life. As a counterpart to the term *trait*, we also want to briefly introduce the term *state*. The current *state* of a person refers to the individual's mood in a specific situation. To clarify this idea, imagine an individual losing a beloved person. Independent of one's own personality traits, clearly suffering from a broken heart will induce a sad condition in (nearly) all humans.

This said, it is also clear that *traits* and *states* interact. It is not as easy to disentangle these terms as it appears in the previous section, because a person who is extraverted (socially outgoing, more positive) is more likely to feel happy at any given time. It has been demonstrated that traits and the means of single state measures over a longer time period correlate between .39 and .64 (Augustine and Larsen 2012). These numbers suggest some overlap of the constructs *traits* and *states*, but also leave enough room for differences. As personality is thought to be stable over time (personality traits change only subtly after the age of 30; McCrae and Costa 1994), it is likely that personality traits represent a vulnerability factor for Internet addiction, rather than that large personality changes are observed after becoming addicted to the Internet. However, given that most studies on Internet addiction deal with younger participants (i.e. below the age of 30), it cannot be ruled out that personality changes may occur due to becoming Internet addicted.

6.3 Is Internet Addiction Heritable? Are Personality and Genetics Related?

By definition genetics can also be viewed as a causal factor for Internet addiction, because genes coding for bodily products clearly represent the starting point for a long cascade of biochemical events, via neurotransmitters, brain structure and functionality, to human behavior, such as being addicted to the Internet. In the following sections we briefly try to answer the question of whether Internet addiction is heritable, and also try to shed some light on the question of how the independent variables of interest in the present chapter—personality and genetics—relate to each other.

The broad topics of the present chapter—personality and genetics—are closely entwined, because personality traits, such as being high/low on anxiety or cooperativeness, are strongly influenced by genetics. Twin studies reported that individual differences in personality show heritability estimates about .50 (Jang et al. 1996). This means that about half of the variation in personality traits between individuals is due to genetic influences. Clearly this represents an oversimplification, because the emerging field of epigenetics demonstrates how environmental influences shape genetic activity (Berger et al. 2009; Nestler 2012). As a consequence, nature and nurture are strongly entwined in shaping psychological phenotypes such as personality traits.

Of note, not only individual differences in personality, but also in the susceptibility to non-substance related addictions, are influenced by genetics, e.g. a review study on pathological gambling by Lobo and Kennedy (2009) illustrates this point nicely showing heritability estimates of .50–.60. Unfortunately, no twin studies on Internet addiction have yet been conducted to allow for estimation of the influence of genetics on the disposition of Internet addiction. Given both the evidence from twin studies on pathological gambling and the findings from molecular genetics

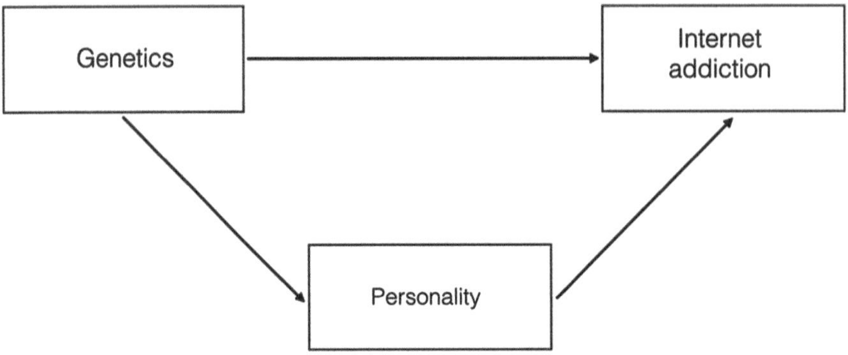

Fig. 6.1 Genetics could influence Internet addiction either directly or mediated via personality traits or other variables

and Internet addiction (see below), it is likely that twin studies in this area will also come up with substantial heritability estimates.[1]

Investigating both molecular genetics and personality in the context of Internet addiction could reflect two sides of the same coin, because genes may exert their influence on Internet addition via personality (e.g. a mediation effect). Clearly it is also imaginable that genetic variations may exert their influence via the non-shared variance of the correlations (e.g. a direct influence) between personality and Internet addiction. This is depicted in Fig. 6.1 and will be discussed later on in this review.

6.4 Personality and Internet Addiction

Before summarizing the large body of studies dealing with Internet addiction and personality, several major issues hindering the generalization of results need to be highlighted.

6.4.1 Problems in Reviewing the 'Personality—Internet Addiction' Findings

First of all, the reviewed studies vary strongly with respect to the personality questionnaires administered. Although personality dimensions can be measured with different questionnaires, the inter-correlations of the different scales are far from

[1] Shortly after completion of the present book, three twin studies on Internet addiction have been published showing that individual differences in Internet addiction may indeed be accounted for by a genetic component (Deryakulu and Ursavas 2014; Li et al. 2014; Vink et al. 2015). In an eventual second edition of this book these studies will be highlighted in more detail.

Table 6.1 A simplified overview on the investigated personality dimensions in the context of Internet addiction

Personality dimension	Explanation
Extraversion (vs. Introversion); can be measured with many questionnaires—e.g. NEO-FFI or NEO-PI-R or EPQ-R	Socially outgoing, optimistic, sometimes impulsive, risk-taking
Neuroticism (vs. Emotional stability); can be measured with many questionnaires—e.g. NEO-FFI or NEO-PI-R or EPQ-R	Emotional instable, anxious, having guilt feeling
Psychoticism can be measured with many questionnaires—e. g. EPQ-R	Cold-hearted, creative, antisocial, egocentric
Novelty seeking (as measured by the Tridimensional Personality Questionnaire or the Temperament and Character Inventory)	Temperament dimension with the subscales exploratory excitability, impulsiveness, extravagance and disorderliness (see also Kose et al. 2003)
Harm avoidance (as measured by the Tridimensional Personality Questionnaire or the Temperament and Character Inventory)	Temperament dimension with the subscales anticipatory worry, fear of uncertainty, shyness and fatigability (see also Kose et al. 2003)
Reward dependence (as measured by the Tridimensional Personality Questionnaire or the Temperament and Character Inventory)	Temperament dimension with the subscales sentimentality, attachment and dependence (see also Kose et al. 2003)
Self-Directedness (the Temperament and Character Inventory)	Character dimension with the subscales responsibility, purposefulness, resourcefulness and congruent second nature (see also Kose et al. 2003)
Impulsivity (as measured by the Barratt Impulsiveness Scale-11)	Impulsivity can be measured with three subscales named attentional, motor and non-planning impulsivity (see also Stanford et al. 2009)

perfect. For example, about 75 % of non-shared variance can be observed when correlating personality traits linked to positive emotionality, such as *Extraversion* and *Novelty Seeking* (Montag et al. 2012). Extraverts can be described as socially outgoing, sometimes impulsive and optimistic. Novelty Seeking is associated with being impulsive, extravagant and loving to explore the world (Cloninger et al. 1993; Kose 2003). For a short overview on the most important personality dimensions in Internet addiction research, please see Table 6.1.

More problems arise when considering further characteristics of the reviewed studies. Some studies clearly only recruited very small samples (e.g. Armstrong et al. 2000, $n = 50$) and most studies focused on young participants, thereby lacking representativeness. In addition, one needs to consider the different cultural backgrounds of the conducted research. Different environments across the globe may have profound influences on human personality (and potentially also on disposition for Internet addiction). Another crucial point when comparing the results of the studies reviewed here concerns the dependent variable of interest, namely Internet addiction. Internet addiction has been measured very differently across all studies, ranging from the inclusion of psychiatrists' assessments, to a range of self-report questionnaires (e.g. Tao et al. 2010; Young 1998a, b). In this context the reader is referred to the important new field of *Psychoinformatics* (Markowetz et al. 2014), which will aid the diagnostic process and treatment of Internet addiction. Please see also the Chap. 9 on this newly emerging topic in this book.

Last but not least, one of the reviewed studies investigated a specific form of Internet addiction—namely online gaming Internet addiction—in the context of personality (Kim et al. 2008). As recently reported by our own work group only generalized Internet addiction and online social network addiction show a large overlap in terms of their inter-correlations (Montag et al. 2014). In contrast, online gaming addiction only shows small correlations with generalized Internet addiction and will therefore be excluded from this summary. Having observed the caveats, the next section will deal with the results of a large number of studies.

6.4.2 Neuroticism, Harm Avoidance, Psychoticism and Internet Addiction

The reviewed studies are all presented in Table 6.2. Summarizing, it becomes obvious that several studies observed higher scores either for the personality trait *Neuroticism* or *Harm Avoidance* in Internet addicts as compared to non-afflicted controls (e.g. Tsai et al. 2009; Hardie and Tee 2007; Cao and Su 2007). *Neuroticism* itself (and also *Harm Avoidance*) is a well-known vulnerability factor for depression and more generally, a public health hazard (Lahey 2009). This also fits with the observation, that Internet addicts or at least a subgroup of patients suffering from Internet addiction show high co-morbidity with affective disorders (e.g. Kim et al. 2006). It is not clear, however, whether depression may cause Internet addiction, or if it represents an outcome of Internet addiction (ultimately, both ways are imaginable and likely).

How can the anxiety-Internet-overuse link be explained? High neurotics/harm avoidant individuals are characterized by constant worrying about the future, and emotional instability. In the context of Internet addiction, being too anxious to interact with others, particularly, in face-to-face interactions, may represent the trigger to overuse the Internet (Shepherd and Edelmann 2005). The online world offers a more anonymous and distant way to fulfill the human need for social interaction, which may be accompanied by social withdrawal from the offline, real world. Clearly, these findings demonstrate that negative emotionality is closely linked to Internet addiction.

Adding to these findings, some studies report an association between *Psychoticism* and Internet addiction (Cao and Su 2007; Xiuqin et al. 2010). Thus, personality characteristics such as cold-heartedness underlying in extreme forms "psychotic and sociopathic behavior" (McCrae and Costa 1985, p. 588) have been linked to excessive Internet use. A look at the findings by Amiel and Sargent (2004) may explain this *Psychoticism* finding from a motivational perspective. Here, persons scoring high on the personality trait of *Psychoticism* showed a higher motivation to use "deviant, defiant, and sophisticated Internet Applications" (p. 711). This includes content such as illegal file sharing and use of pornography.

Table 6.2 Studies on personality and Internet addiction (some of the studies focussed also on other aspects besides personality, which are not summarized in the present review)

Authors	Participants	Internet addiction and personality inventories, etc.	Results
Armstrong et al. (2000)	N = 50 participants	Coopersmith Self-Esteem Inventory, Sensation Seeking Scale, MMPI-2 Addiction Potential Scale, Internet Related Problem Scale	Low *self-esteem* was associated with Internet addiction. No association between *Impulsivity* (measured via the disinhibition scale of *Sensation Seeking*) could be observed with Internet addiction
Cao et al. (2007)	N = 2620 high school students from which a case-control study was conducted (with 64 Internet addicts and 64 control persons)	Eysenck Personality Questionnaire (edition for children), diagnostic questionnaire for Internet addiction	Internet addicts were associated with higher *Neuroticism*, higher *Psychoticism* and *Lie scores*
Cao et al. (2007)	N = 50 Internet addicts were contrasted with N = 50 healthy Internet users	Barratt Impulsivity Scale; diagnostic questionnaire for Internet addiction	Internet addicts showed higher *Impulsivity* scores than controls
Ha et al. (2007)	N = 452 Korean adolescents	Temperament and Character Inventory, Internet Addiction Test	Internet addiction was associated with high *Harm Avoidance*, low *Self-Directedness*, low *Cooperativeness* and high *Self-Transcendence*
Han et al. (2007)	N = 154 participants (79 Internet addicts and 75 healthy controls)	Temperament and Character Inventory and Internet Addiction Test	Higher *Reward Dependence* in Internet addicts compared to healthy control persons
Hardie and Tee (2007)	N = 96 participants	International Personality Item Pool, Internet Addiction Test	*Neuroticism* is associated with Internet addiction
Kim et al. (2008)	N = 1471 online game users	Buss-Perry Aggression Questionnaire, Self-Control Scale, Narcissistic Personality Disorder Scale, Online Game Addiction Scale (modified from Young's Internet Addiction Test)	*Aggression* and *narcissistic* personality was positively associated with online game addiction. Moreover a negative association with self-control turned up.

(continued)

Table 6.2 (continued)

Authors	Participants	Internet addiction and personality inventories, etc.	Results
Ko et al. (2010)	N = 216 college students	Symptom criteria were used for diagnosing Internet addiction, Tridimensional Personality Questionnaire (TPQ)	Lower *Reward Dependence* and higher *Harm Avoidance* scores for Internet addicts
Ko et al. (2006)	N = 3662 students	Chen Internet Addiction Scale, TPQ	High *Novelty Seeking*, high *Harm Avoidance* and low *Reward Dependence* predicted Internet addiction
Lee et al. (2012)	N = 27 Internet addicts; N = 27 patients with pathological gambling, N = 27 healthy controls	Barratt Impulsiveness Scale, Young's Internet Addiction Test	Internet addicts show significantly higher *Impulsivity* scores than control persons (and comparable scores to the group of pathological gamblers)
Montag et al. (2010)	N = 201 participants	Eysenck's Personality Inventory (-R), Revised, Temperament and Character Inventory, Internet Addiction Test	Low *Self-Directedness* is a better predictor for Internet addiction than *Neuroticism* or *Harm Avoidance*
Montag et al. (2011)	N = 610 participants	Among others NEO-FFI, Self-Directedness of the Temperament and Character Inventory and Internet Addiction Test	Low *Conscientiousness* and low *Self-Directedness* are associated with higher Internet addiction
Mottram and Fleming (2009)	N = 272 undergraduate students	UPPS Impulsive Behavior Scale, Extraversion from the Big Five Inventor	With respect to personality 'lack of perseverance' (a facet of impulsivity) was a predictor for Internet addiction
Sariyska et al. (2014)	N = 989 participants from seven countries	Self-Directedness of the Temperament and Character Inventory and Internet Addiction Test	In all countries under investigation (China, Taiwan, Germany, Bulgaria, Spain, Sweden and Colombia) a negative association between high *Self-Directedness* and low Internet addiction could be observed

(continued)

Table 6.2 (continued)

Authors	Participants	Internet addiction and personality inventories, etc.	Results
Tsai et al. (2009)	N = 1360 freshman in Taiwan	Chinese Internet Addiction Scale (-R), Chinese Health Questionnaire, Maudsely Personality Inventory	Male gender, *Neuroticism* and higher probability for psychiatric disorders are associated with Internet addiction
Yan et al. (2014)	N = 892 college students	Eysenck's Personality Questionnaire, Chen Internet Addiction Scale	Among others Internet addicts were characterized by higher *Neuroticism*, lower *Extraversion* and higher *Psychoticism*
Yen et al. (2009)	N = 1992 college students	Behavioral Inhibition/Activation Scale (BIS/BAS), Chen Internet Addiction Scale	Higher scores of *BIS* and the *BAS Fun Seeking* subscale in Internet addicts
Xiuqin et al. (2010)	N = 204 Internet addicts and N = 100 controls	Eysenck Personality Questionnaire Revised, Symptom Checklist 90 Revision, My Memories of Upbringing Scale, Internet addiction diagnosed via symptoms	Internet addicts are associated with lower *Extraversion* and higher *Psychoticism* scores. Moreover Internet addicts reported more symptoms on the symptom checklist 90 and less parental support

Following this, *Psychoticism* may be of special, and greater relevance when investigating specific forms of IA, such as online pornography addiction, rather than generalized Internet addiction. Finally, it has recently been suggested that high *Psychoticism* together with high life stress may represent a unique vulnerability constellation for Internet addiction (Yan et al. 2014). Of note, personality traits such as *Psychoticism* need to be seen in the context of a continuum model. This means that humans can be characterized by lower or higher scores on such a personality dimension, but not in terms of distinct categories such as being a psychopath or not. As a consequence psychotic behavior is not per se psychopathological.

Of note, further studies have tried to link Cloninger's character dimension *Reward Dependence* (Han et al. 2007; Ko et al. 2006), as well as individual differences in *Impulsivity* (Armstrong et al. 2000; Cao et al. 2007; Lee et al. 2012), to Internet addiction. As described in Table 6.1, humans scoring high on *Reward Dependence* are sentimental, attach easily to others and give a lot of attention to what others think about them. Impulsivity can be viewed as "swift action without forethought or conscious judgment" (Moeller et al. 2001, p. 1783). Impulsive behavior can be measured by the Disinhibition scale of the construct *Sensation Seeking* by Zuckerman (1990) or the Barratt Impulsiveness Scale (e.g. Stanford et al. 2009).

The studies dealing with *Reward Dependence* are inconsistent, as the results of the studies by Ko et al. (2006) and Han et al. (2007) point in opposite directions. The results on impulsive behavior are a bit more coherent (in particular those using the Barratt Impulsivity scale), linking high impulsivity to Internet addiction. This link is also supported by academic discussions in favour of characterizing Internet addiction as an impulse control disorder (Shapira et al. 2003). Moreover, an association between the attention deficit and hyperactivity disorder (ADHD) and Internet addiction has been already established; which is particularly interesting as ADHD is also accompanied by impulsive behavior (Yen et al. 2007).

6.4.3 Individual Differences in Self-Directedness as a Core Feature of Internet Addiction?

The question arises, if traits related to negative emotionality and *Psychoticism* represent the best predictors for (generalized) Internet addiction. Our own data suggest that this might only partially be the case. Rather, a series of our own studies (Montag et al. 2010, 2011; Sariyska et al. 2014) showed that individual differences in the character dimension *Self-Directedness* may be a better predictor for Internet addiction. People scoring high on *Self-Directedness* are satisfied with their personality, accept themselves as they are and handle their everyday life successfully. Clearly, these persons are also characterized by high self-esteem and high willpower.

In an initial study, we demonstrated that low *Self-Directedness* represents a better predictor for Internet addiction than high *Neuroticism/Harm Avoidance* (and *Psychoticism*, which played no role). In detail, the variables *hours of leisure Internet use each week* together with (low) *Self-Directedness* explained 25 % of the variance in Internet addiction (Montag et al. 2010). *Harm Avoidance* and *Neuroticism* did not explain any additional variance in Internet addiction, when the two aforementioned variables were included in the regression model.

As this study was conducted with a student sample not at high risk for Internet addiction, we tried to replicate the link between low *Self-Directedness* and high Internet addiction in more than 600 participants, who were characterized by being online video gamers of first-person-shooter-video-games (Montag et al. 2011). Although online video gamers focus on online games and are more prone to become addicted to this specific form of Internet addiction, they are also characterized by higher generalized Internet addiction (Montag et al. 2011). The association between *Self-Directedness* and Internet addiction was even stronger in this follow up study [compare $r = -.48$, $p < 001$ in the Montag et al. (2011) study to $r = -.35$, $p < .001$ in the above mentioned Montag et al. (2010) study], thereby giving support for a continuum model to explain the 'Self-Directedness Internet addiction link' from healthy to psychopathological use of the Internet. In addition, high *Conscientiousness* was negatively correlated with Internet addiction in this study.

Strongest support for the relevance of *Self-Directedness* in the context of Internet addiction came recently from a cross-cultural study including samples from China, Taiwan, Bulgaria, Spain, Sweden, Germany and Colombia (Sariyska et al. 2014). Despite differences in sociodemographic variables such as age and gender, and of course in cultural background across samples, low *Self-Directedness* was robustly associated with high Internet addiction scores in every sample under investigation. The importance of *Self-Directedness* has also been outlined in broader clinical terms by Cloninger et al. (1993), who state that *Self-Directedness* "is the major determinant of the presence or absence of personality disorder" (p. 979). Although we are not dealing with personality disorders in the present chapter, Cloninger et al. hint towards the tremendous clinical relevance of *Self-Directedness*, which is reflected by our present Internet addiction studies, too.

In sum, we believe that the character dimension *Self-Directedness*, of Cloninger's Temperament and Character Inventory, represents a valuable tool in the treatment and research of Internet addiction. Deriving from our *Self-Directedness* results, therapeutic treatment of Internet addicts should aim at restructuring the patients' lives. Moreover, the therapy should focus on the acceptance of one's own strength and weaknesses, and on the related area of self-esteem.

6.5 Molecular Genetics and Internet Addiction

As mentioned in the introduction, the present chapter also deals with molecular genetics and Internet addiction. To date, three studies have investigated—by means of a candidate gene approach—the molecular genetics of Internet addiction.

The first study was published by Han et al. (2007), and investigated two prominent dopaminergic genetic variations in the context of Internet addiction. A role for dopamine in Internet addiction is very probable, because reduced dopamine transporters have been observed in striatal regions of the human (Internet addicted) brain (Hou et al. 2012; Kim et al. 2011) and Internet gaming addicts show stronger striatal responses to gaming cues compared with controls (Ko et al. 2009). In general, dopamine represents the classic neurotransmitter in the investigation of addiction because it is discussed to be the final common pathway of reward (Pierce and Kumaresan 2006). The molecular genetic findings by Han et al. (2007) provide support for a role for dopamine in Internet addiction, as the catechol-o-methyltransferase (COMT) Val158Met polymorphism, which crucially influences the dopamine catabolism, was associated with Internet addiction. The COMT 158Met allele variant, associated with a 3–4 fold lower activity of the COMT enzyme (Lachman et al. 1996), occurred significantly more often in Internet addicts compared to controls. In addition to this dopaminergic target, the authors also investigated the ANKK1/DRD2 Taq Ia polymorphism, which is associated with individual differences in D2 receptor density in striatal regions (Pohjalainen et al. 1998). Here, Internet addicts were linked to a higher occurrence of the A1+ variant (carrying at least one A1 allele) compared to controls. Of note the A1+ variant is associated with a 30–40 % reduction in D2 receptors (Pohjalainen et al. 1998), and has previously been associated with alcoholism (Munafo et al. 2007). Other addictive forms including opioids or smoking have been also associated with this genetic variant (e.g. De Ruyck et al. 2010; Wang et al. 2013). Deriving from this, similar biochemical processes could underlie substance and non-substance related addictions.

The Han et al. (2007) study also reported that Internet addicts are associated with higher *Reward Dependence* scores. In addition, among Internet addicts, carriers of the A1+ variant showed significantly higher scores in *Reward Dependence* compared to control persons. This provides support for the earlier suggestion, depicted in Fig. 6.1, that genetic variations may target personality traits and thereby indirectly influence vulnerability to Internet addiction. Although this was not tested directly in the Han et al. study, the results do point in this direction.

Further indirect support for a role of dopamine in Internet addiction comes from a study investigating the CHRNA4 gene coding for the subunit alpha 4 of the nicotinic acetylcholine receptor gene (Montag et al. 2012). Of special note for the dopamine link is the observation that the administration of nicotine, such as by smoking a cigarette, triggers the dopaminergic system (Corrigall et al. 1992). In the context of the investigated CHRNA4 gene, carriers of the CC variant of the single nucleotide polymorphism rs1044396 on this gene were associated with Internet addiction, because this genetic variation occurred significantly more often in the Internet addict group compared to the control group. Interestingly, this same CC variant associated with Internet addiction has previously also been associated with smoking (Feng et al. 2004).

A further statistical analysis of the CHRNA4-Internet addiction link revealed that this effect was mainly observed in female Internet addicts. This could be explained by the potentially higher use of social networking sites in females (Thelwall 2008), although this statistical effect could be rather small, and may depend on the site(s) used (Hargittai 2007). Data on the gender-social network link could not be provided in the Montag et al. (2012) study, because activities in special domains of the Internet use were not collected.

Besides dopamine, mounting evidence from psychopharmacology and molecular genetics suggests that serotonin may also play a crucial role in the biochemistry of Internet addiction. First evidence from the psychopharmacological treatment of Internet addicts shows some success in treating Internet addicts with selective serotonin reuptake inhibitors (e.g. Atmaca 2007; Camardese et al. 2012, see Chap. 10). Moreover, Lee et al. (2008) demonstrated that a genetic variation of the gene SLC6A4 coding for the serotonin transporter, could be associated with Internet addiction. In detail, the ss-variant of the so-called serotonin transporter polymorphism 5-HTTLPR occurred more often in Internet addicts compared to control persons. Of note, the s-variant has been associated with lower mRNA expression of the serotonin transporter gene and therefore, putatively, also with a lower number of presynaptic transporters (Lesch et al. 1996). It has to be pointed out that 5-HTTLPR has meanwhile qualified as a super-vulnerability-factor for psychopathological disorders. Since Lesch et al. demonstrated for the first time in 1996 an association between 5-HTTLPR and the personality factor of neuroticism, countless studies have reported associations between this polymorphism and all sorts of disorders, among these also drug addiction (for a review see Kenna et al. 2012).

6.6 General Internet Usage and Personality

The present chapter focused explicitly on a review of studies dealing with Internet addiction, molecular genetics and personality. There are several studies dealing also with a description of the link between normal/healthy use of the Internet and personality, which has not been considered in the present chapter, e.g. Hamburger and Ben-Artzi (2000) reported that male extraverts use the Internet for leisure activities, whereas extraversion was negatively correlated with (online) leisure activities in females. Moreover, female neurotics more often used online social services. Interestingly, Hills and Argyle (2003) could not find a strong link between the use of 16 different services and individual differences in personality when age and gender were controlled for. This is not surprising, because Hamburger and Ben-Artzi (2000) provided evidence for gender dependent differences in Internet usage. An interesting observation has been made by Amiel and Sargent (2004) while researching motives for Internet usage. Neurotics used the Internet to fulfill a need for belonging and being informed. Extraverts used the Internet in very goal oriented manners (e.g. research or music sharing) but not for reasons of meeting people online. Finally, users scoring high on *Psychoticism* showed interest in more extreme Internet usage (as mentioned above).

6.7 Conclusions

The present review showed that genetic variations of the dopaminergic, serotonergic and cholinergic system are associated with Internet addiction. From classic personality psychology comes evidence that personality traits linked to negative emotionality, but perhaps more important to self-regulation and self-esteem, are strongly linked to Internet addiction. As both the genetic variations and personality traits discussed here are known to play a role in other addictions (e.g. Basiaux et al. 2001; Guo et al. 2007), it is likely that same (bio)psychological mechanisms underlie Internet addiction as other addictions.

References

Amiel T, Sargent SL (2004) Individual differences in internet usage motives. Comput Hum Behav 20:711–726

Armstrong L, Phillips JG, Saling LL (2000) Potential determinants of heavier internet usage. Int J Hum Comput Stud 53:537–550

Atmaca M (2007) A case of problematic Internet use successfully treated with an SSRI-antipsychotic combination. Prog Neuropsychopharmacol Biol Psychiatry 31:961–962

Augustine AA, Larsen RJ (2012) Is a trait really the mean of states? Similarities and differences between traditional and aggregate assessments of personality. J Individ Differ 33:131

Basiaux P, Le Bon O, Dramaix M et al (2001) Temperament and Character Inventory (TCI) personality profile and sub-typing in alcoholic patients: a controlled study. Alcohol Alcohol 36:584–587

Berger SL, Kouzarides T, Shiekhattar R, Shilatifard A (2009) An operational definition of epigenetics. Genes Dev 23:781–783

Camardese G, De Risio L, Di Nicola M et al (2012) A role for pharmacotherapy in the treatment of "internet addiction". Clin Neuropharmacol 35:283–289

Cao FL, Su LY (2007) Internet addiction among Chinese adolescents: prevalence and psychological features. Child Care Health Dev 33:275–281

Cao F, Su L, Liu T, Gao X (2007) The relationship between impulsivity and internet addiction in a sample of Chinese adolescents. Eur Psychiatry 22:466–471

Chak K, Leung L (2004) Shyness and locus of control as predictors of internet addiction and internet use. Cyberpsychol Behav 7:559–570

Cloninger CR, Svrakic DM, Przybeck TR (1993) A psychobiological model of temperament and character. Arch Gen Psychiatry 50:975

Corrigall WA, Franklin KB, Coen KM, Clarke PB (1992) The mesolimbic dopaminergic system is implicated in the reinforcing effects of nicotine. Psychopharmacology 107:285–289

De Ruyck K, Nackaerts K, Beels L et al (2010) Genetic variation in three candidate genes and nicotine dependence, withdrawal and smoking cessation in hospitalized patients. Pharmacogenomics 11:1053–1063

Deryakulu D, Ursavaş ÖF (2014) Genetic and environmental influences on problematic Internet use: a twin study. Comput Hum Behav 39:331–338

Dong G, Lu Q, Zhou H, Zhao X (2010) Impulse inhibition in people with Internet addiction disorder: electrophysiological evidence from a Go/NoGo study. Neurosci Lett 485:138–142

Feng Y, Niu T, Xing H et al (2004) A common haplotype of the nicotine acetylcholine receptor α4 subunit gene is associated with vulnerability to nicotine addiction in men. Am J Human Genet 75:112–121

Guo S, Zhou DF, Sun HQ et al (2007) Association of functional catechol O-methyl transferase (COMT) Val108Met polymorphism with smoking severity and age of smoking initiation in Chinese male smokers. Psychopharmacology 190:449–456

Ha JH, Kim SY, Bae SC et al (2007) Depression and internet addiction in adolescents. Psychopathology 40:424–430
Hamburger YA, Ben-Artzi E (2000) The relationship between extraversion and neuroticism and the different uses of the internet. Comput Hum Behav 16:441–449
Han DH, Lee YS, Yang KC et al (2007) Dopamine genes and reward dependence in adolescents with excessive internet video game play. J Addict Med 1:133–138
Hardie E, Tee MY (2007) Excessive internet use: the role of personality, loneliness and social support networks in internet addiction. Aust J Emerg Technol Soc 5:34–47
Hargittai E (2007) Whose space? Differences among users and non-users of social network sites. J Comput Mediated Commun 13:276–297
Hills P, Argyle M (2003) Uses of the internet and their relationships with individual differences in personality. Comput Hum Behav 19:59–70
Hou H, Jia S, Hu S et al. (2012) Reduced striatal dopamine transporters in people with internet addiction disorder. BioMed Res Int 854524. doi: 10.1155/2012/854524
Hur MH (2006) Demographic, habitual, and socioeconomic determinants of internet addiction disorder: an empirical study of Korean teenagers. Cyberpsychol Behav 9:514–525
Jang KL, Livesley WJ, Vemon PA (1996) Heritability of the big five personality dimensions and their facets: a twin study. J Pers 64:577–592
Kenna GA, Roder-Hanna N, Leggio L et al (2012) Association of the 5-HTT gene-linked promoter region (5-HTTLPR) polymorphism with psychiatric disorders: review of psychopathology and pharmacotherapy. Pharmacogenomics Personalized Med 5:19–35
Kim K, Ryu E, Chon MY et al (2006) Internet addiction in Korean adolescents and its relation to depression and suicidal ideation: a questionnaire survey. Int J Nurs Stud 43:185–192
Kim EJ, Namkoong K, Ku T, Kim SJ (2008) The relationship between online game addiction and aggression, self-control and narcissistic personality traits. Eur Psychiatry 23:212–218
Kim SH, Baik SH, Park CS et al (2011) Reduced striatal dopamine D2 receptors in people with internet addiction. NeuroReport 22:407–411
Ko CH, Yen JY, Chen CC et al (2006) Tridimensional personality of adolescents with internet addiction and substance use experience. Can J Psychiatry Rev Can Psychiatr 51:887–894
Ko CH, Liu GC, Hsiao S et al (2009) Brain activities associated with gaming urge of online gaming addiction. J Psychiatr Res 43:739–747
Ko CH, Hsiao S, Liu GC, Yen JY, Yang MJ, Yen CF (2010) The characteristics of decision making, potential to take risks, and personality of college students with Internet addiction. Psychiatry Res\175(1):121–125
Ko CH, Yen JY, Yen CF et al (2012) The association between internet addiction and psychiatric disorder: a review of the literature. Eur Psychiatry 27:1–8
Kose (2003) Psychobiological model of temperament and character: TCI. Yeni Symp 41(2):86–97
Lachman HM, Papolos DF, Saito T et al (1996) Human catechol-O-methyltransferase pharmacogenetics: description of a functional polymorphism and its potential application to neuropsychiatric disorders. Pharmacogenet Genomics 6:243–250
Lahey BB (2009) Public health significance of neuroticism. Am Psychol 64:241
Lee YS, Han DH, Yang KC et al (2008) Depression like characteristics of 5HTTLPR polymorphism and temperament in excessive internet users. J Affect Disord 109:165–169
Lee HW, Choi JS, Shin YC et al (2012) Impulsivity in internet addiction: a comparison with pathological gambling. Cyberpsychol Behav Soc Networking 15:373–377
Lesch KP, Bengel D, Heils A et al (1996) Association of anxiety-related traits with a polymorphism in the serotonin transporter gene regulatory region. Science 274:1527–1531
Li M, Chen J, Li N, Li X (2014) A twin study of problematic internet use: its heritability and genetic association with effortful control. Twin Res Hum Genet 17(04),279–287
Lobo DS, Kennedy JL (2009) Genetic aspects of pathological gambling: a complex disorder with shared genetic vulnerabilities. Addiction 104:1454–1465
Markowetz A, Błaszkiewicz K, Montag C et al (2014) Psycho-informatics: big data shaping modern psychometrics. Med Hypotheses 82:405–411
McCrae RR, Costa PT Jr (1985) Comparison of EPI and psychoticism scales with measures of the five-factor model of personality. Personality Individ Differ 6:587–597

McCrae RR, Costa PT (1994) The stability of personality: observation and evaluations. Curr Dir Psychol Sci 3:173–175
Moeller FG, Barratt ES, Dougherty DM et al (2001) Psychiatric aspects of impulsivity. Am J Psychiatry 158:1783–1793
Montag C, Jurkiewicz M, Reuter M (2010) Low self-directedness is a better predictor for problematic internet use than high neuroticism. Comput Hum Behav 26:1531–1535
Montag C, Flierl M, Markett S et al (2011) Internet addiction and personality in first-person-shooter video gamers. J Media Psychol Theor Methods Appl 23:163
Montag C, Jurkiewicz M, Reuter M (2012a) The role of the catechol-O-methyltransferase (COMT) gene in personality and related psychopathological disorders. CNS Neurol Dis Drug Targets (Formerly Curr Drug Targets CNS Neurol Dis) 11:236–250
Montag C, Kirsch P, Sauer C et al (2012b) The role of the CHRNA4 gene in internet addiction: a case-control Study. J Addict Med 6:191–195
Montag C, Reuter M, Jurkiewicz M et al (2013) Imaging the structure of the human anxious brain: a review of findings from neuroscientific personality psychology. Rev Neurosci 24:167–190
Montag C, Bey K, Sha P et al (2014) Is it meaningful to distinguish between generalized and specific Internet addiction? Evidence from a cross-cultural study from Germany, Sweden. Asia-Pacific Psychiatry, Taiwan. doi:10.1111/appy.12122
Mottram AJ, Fleming MJ (2009) Extraversion, impulsivity, and online group membership as predictors of problematic internet use. Cyberpsychol Behav 12:319–321
Munafo MR, Matheson IJ, Flint J (2007) Association of the DRD2 gene Taq1A polymorphism and alcoholism: a meta-analysis of case–control studies and evidence of publication bias. Mol Psychiatry 12:454–461
Nestler EJ (2012) Epigenetics: stress makes its molecular mark. Nature 490:171–172
Pierce RC, Kumaresan V (2006) The mesolimbic dopamine system: the final common pathway for the reinforcing effect of drugs of abuse? Neurosci Biobehav Rev 30:215–238
Pohjalainen T, Rinne JO, Någren K et al (1998) The A1 allele of the human D2 dopamine receptor gene predicts low D2 receptor availability in healthy volunteers. Mol Psychiatry 3:256–260
Rumpf HJ, Meyer C, Kreuzer A et al. (2011) Prävalenz der Internetabhängigkeit. Bericht an das Bundesministerium für Gesundheit. Available via. http://www.drogenbeauftragte.de/presse/pressemitteilungen/2011-03/pinta-studie.html. Accessed 24 Jan 2012
Sariyska R, Reuter M, Bey K et al (2014) Self-esteem, personality and internet addiction: a cross-cultural comparison study. Personality Individ Differ 61:28–33
Shapira NA, Lessig MC, Goldsmith TD et al (2003) Problematic internet use: proposed classification and diagnostic criteria. Depress Anxiety 17:207–216
Shaw M, Black DW (2008) Internet addiction. CNS Drugs 22:353–365
Shepherd RM, Edelmann RJ (2005) Reasons for internet use and social anxiety. Personality Individ Differ 39:949–958
Stanford MS, Mathias CW, Dougherty DM et al (2009) Fifty years of the Barratt impulsiveness scale: an update and review. Personality Individ Differ 47:385–395
Tao R, Huang X, Wang J et al (2010) Proposed diagnostic criteria for internet addiction. Addiction 105:556–564
Thelwall M (2008) Social networks, gender, and friending: an analysis of MySpace member profiles. J Am Soc Inform Sci Technol 59:1321–1330
Tsai HF, Cheng SH, Yeh TL et al (2009) The risk factors of Internet addiction—a survey of university freshmen. Psychiatry Res 167:294–299
Vink JM, Beijsterveldt TC, Huppertz C, Bartels M, Boomsma DI (2015) Heritability of compulsive Internet use in adolescents. Addiction biol. doi:10.1111/adb.12218
Wang TY, Lee SY, Chen SL et al (2013) Association between *DRD2 5-HTTLPR* and *ALDH2* genes and specific personality traits in alcohol-and opiate-dependent patients. Behav Brain Res 250:285–292
Xiuqin H, Huimin Z, Mengchen L et al (2010) Mental health, personality, and parental rearing styles of adolescents with internet addiction disorder. Cyberpsychol Behav Social Networking 13:401–406

Yan W, Li Y, Sui N (2014) The relationship between recent stressful life events, personality traits, perceived family functioning and internet addiction among college students. Stress Health 30(1):3–11
Yen JY, Ko CH, Yen CF et al (2007) The comorbid psychiatric symptoms of internet addiction: attention deficit and hyperactivity disorder (ADHD), depression, social phobia, and hostility. J Adolesc Health 41:93–98
Yen JY, Ko CH, Yen CF, Chen CS, Chen CC (2009) The association between harmful alcohol use and Internet addiction among college students: comparison of personality. Psychiatry Clin Neurosci 63(2):218–224
Young KS (1998a) Internet addiction: the emergence of a new clinical disorder. Cyberpsychology Behav 1:237–244
Young KS (1998b) Caught in the net: how to recognize the signs of internet addiction—and a winning strategy for recovery. Wiley, Hoboken
Yuan K, Qin W, Wang G et al (2011) Microstructure abnormalities in adolescents with internet addiction disorder. PLoS ONE 6:e20708
Zhou Y, Lin FC, Du YS et al (2011) Gray matter abnormalities in Internet addiction: a voxel-based morphometry study. Eur J Radiol 79:92–95
Zuckerman M (1990) The psychophysiology of sensation seeking. J Pers 58:313–345

Chapter 7
Autonomic Nervous System and Brain Circuitry for Internet Addiction

Andrew Chih Wei Huang

Abstract The autonomic nervous system can be divided into sympathetic and parasympathetic divisions. The activation of the sympathetic and parasympathetic autonomic nervous systems in Internet addiction is similar to that in drug addiction. The sympathetic division is more strongly reactive, with responses in internal glands and bodily organs and acceleration of heart rate (or blood volume pulse) and respiratory response rate as well as a reduction of temperature, while surfing the Internet. However, contradictory data have also been obtained. Decelerations of skin conductance have also been observed in Internet addiction, showing activity of the parasympathetic division, but not the sympathetic division. Drug addiction produces greater activation of the sympathetic nervous system, but less activation of the parasympathetic system. The autonomic nervous system reciprocally connects to neural circuitry in the brain to modulate both systems, reflecting specific features of Internet addiction. Moreover, Internet addiction induces numerous behavioral and psychopathological symptoms related to pathogenesis, including depression, anxiety, hostility, psychoticism, interpersonal sensitivity, attention-deficit/hyperactivity disorder, obsessive-compulsive disorder, novelty seeking, and social anxiety disorder. Therapeutic interventions for Internet addiction, including pharmacological and non-pharmacological treatments, need to be linked to psychopathological symptoms. The autonomic nervous system, brain circuitry pathogenesis, and specific interventions targeting Internet addiction are discussed in the present review article.

7.1 Introduction

Internet addiction is a newly emerging and increasing prevalent addictive disorder. In contrast to drug addiction, Internet addiction does not directly act via substance consumption on receptors of neural substrates to induce compulsive and impulsive

---

A.C.W. Huang (✉)
Department of Psychology, Fo Guang University, Yi-Lan, Taiwan
e-mail: acwhuang@gmail.com

C. Montag and M. Reuter (eds.), *Internet Addiction*, Studies in Neuroscience, Psychology and Behavioral Economics, DOI 10.1007/978-3-319-07242-5_7

Table 7.1 The neural mechanisms, pathogenesis, and interventions of Internet addiction

Internet addiction		
Neural mechanisms	Pathogenesis	Interventions
1. Central nervous system: neural substrates	Psychopathological behaviors/symptoms	1. Pharmacological treatments
2. Peripheral nervous system: autonomic nervous system (1) Sympathetic nervous system (2) Parasympathetic nervous system	1. Before Internet addiction: Obsessive-compulsive disorder	2. Non-pharmacological treatments
	2. After Internet addiction: Depression, anxiety, hostility, interpersonal sensitivity, and psychoticism	

behavior. Thus, while Internet addiction may partially share neural substrates with drug addiction, it also involves some different brain mechanisms to drug addiction. Comparisons of neural substrates in the peripheral autonomic and central nervous systems of the brain between Internet addiction and drug addiction allows a better understanding of the behavioral and neural mechanisms of Internet addiction.

Internet addiction has been linked to many psychopathological and behavioral symptoms, such as depression, anxiety, hostility, psychoticism, interpersonal sensitivity, attention-deficit/hyperactivity disorder, obsessive-compulsive disorder, novelty seeking, and social anxiety disorder. The pathogenesis of Internet addiction is discussed in the present article, and the current discussion may provide some novel viewpoints into therapeutic interventions for Internet addiction.

To enable a better understanding of Internet addiction, it is important to combine our knowledge of the activity of the autonomic nervous system and neural circuitry in the brain, as the pathogenesis of Internet addiction presumably results from behavioral and neural activity arising from these systems. Accordingly, interventions for Internet addiction may be based on the pathogenesis of Internet addiction. Therefore, three components of Internet addition—pathogenesis, neural mechanisms, and interventions—are interconnected, and their relationship is shown in Table 7.1. The present review article discusses these three aspects of Internet addiction.

7.2 Internet Addiction and the Peripheral Nervous System

7.2.1 Internet Addiction Impacts the Autonomic Nervous System

To our knowledge, little research has investigated how Internet addiction affects the autonomic nervous system with regard to the sympathetic and parasympathetic divisions. Our 2010 study was the first to delve into this interesting issue. The study used the Chen Internet Addiction Scale (CIAS) to assess the magnitude of Internet addiction and to screen participants for allocation into low- and high-risk Internet abuse groups.

Table 7.2 Spearman correlations among BVP, SC, TEMP, RESPR, and CIAS score

	BVP	SC	TEMP	RESPR	CIAS score
BVP (%) (Mean = 0.024, SE = 0.011)	1.000				
SC (%) (Mean = 0.649, SE = 0.223)	−0.101 0.239	**1.000**			
TEMP (%) (Mean = 0.002, SE = 0.009)	−0.056 0.346	**−0.461**** **0.000**	1.000		
RESPR (%) (Mean = 0.202, SE = 0.039)	−0.044 0.379	**−0.238*** **0.045**	0.001 0.497	1.000	
CIAS score (Mean = 56.221, SE = 1.605)	0.188 0.091	**−0.065** **0.323**	−0.312* 0.012	0.336** 0.007	1.000

**$p < 0.01$, *$p < 0.05$, significant α value of Spearman correlation

Four psychophysiological assessments of autonomic nervous activity, including blood volume pulse (BVP), skin conductance (SC), peripheral temperature (TEMP), and respiratory response (RESPR), were recorded while the participants spent 6 min browsing the Internet. Some important findings were obtained in this study. First, RESPR and TEMP were sensitive psychophysiological indices of Internet addiction that were positively and negatively correlated with CIAS score, respectively. Blood volume pulse and SC were unrelated to CIAS score (Table 7.2). Blood volume pulse and RESPR in high-risk Internet abusers were significantly increased compared with low-risk Internet abusers. Skin conductance and TEMP were lower in high-risk abusers compared with low-risk abusers (Figs. 7.1, 7.2, 7.3 and 7.4). This means that greater activation of the sympathetic nervous system, reflected in the increased BVP and RESPR together with decreased TEMP, was observed in the high-risk Internet abusers. However, the SC results showed paradoxical responses in comparison to the other psychophysiological indices. Lower skin conductance activation was observed among high-risk abusers, suggesting that the parasympathetic nervous system was simultaneously activated in this group. Based on these observations, the autonomic activity hypothesis of Internet addiction is suggested to explain how the sympathetic and parasympathetic divisions of the autonomic nervous system are activated in Internet abusers.

The autonomic nervous system is composed of sympathetic and parasympathetic divisions (Fig. 7.5). With regard to structure, the sympathetic and parasympathetic nervous systems can be divided into preganglionic and postganglionic neurons. The dissociation between the sympathetic and parasympathetic divisions involves preganglionic fibers in the sympathetic nervous system that are shorter than those in the parasympathetic nervous system. Similarly, preganglionic fibers in the parasympathetic nervous system are longer than those in the sympathetic nervous system. With regard to functional aspects, the sympathetic nervous system has been shown to govern and enhance the activation of internal glands and bodily organs. The parasympathetic nervous system plays an inhibitory role to attenuate the activity of glands and organs. The sympathetic/parasympathetic divisions are activated within the same internal glands and bodily organs in a complementary way. When the sympathetic division is

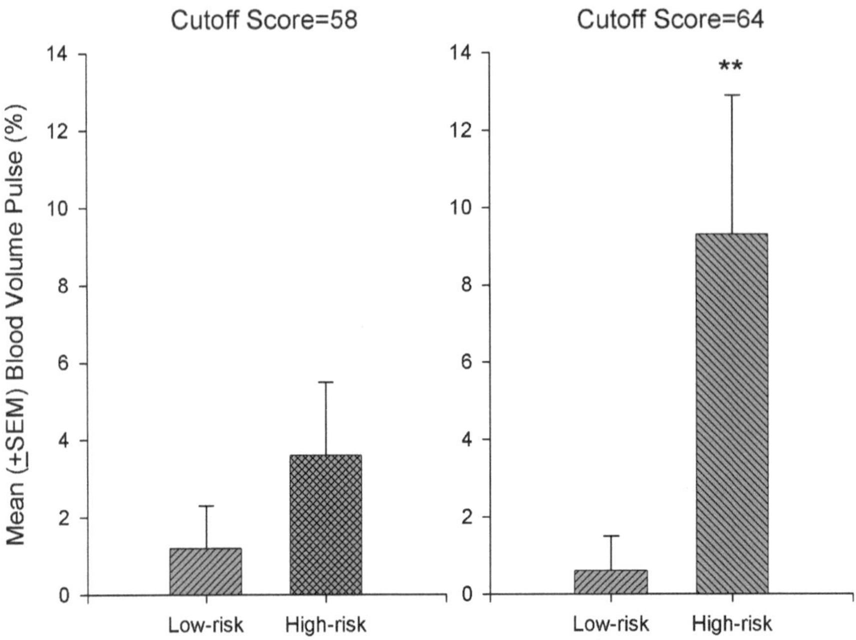

Fig. 7.1 Mean (± SEM) BVP (%) in high- and low-risk Internet abusers, with cutoff scores of 58 or 64 in the CIAS screening of Internet addiction. $^{**}p < 0.01$, versus low-risk group with the same cutoff score

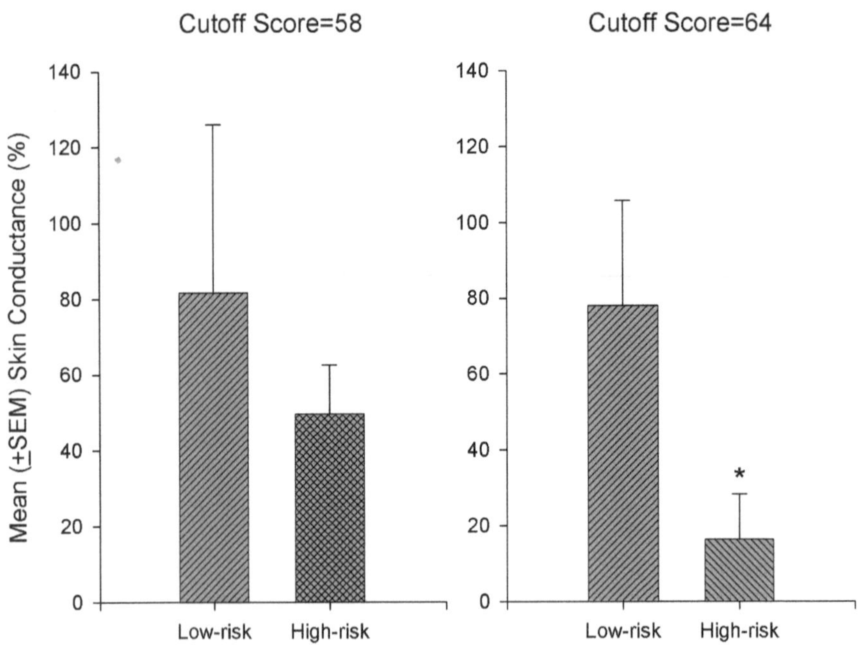

Fig. 7.2 Mean (± SEM) SC (%) in high- and low-risk Internet abusers, with cutoff scores of 58 or 64 in the CIAS screening of Internet addiction. $^{*}p < 0.05$, versus low-risk group with the same cutoff score

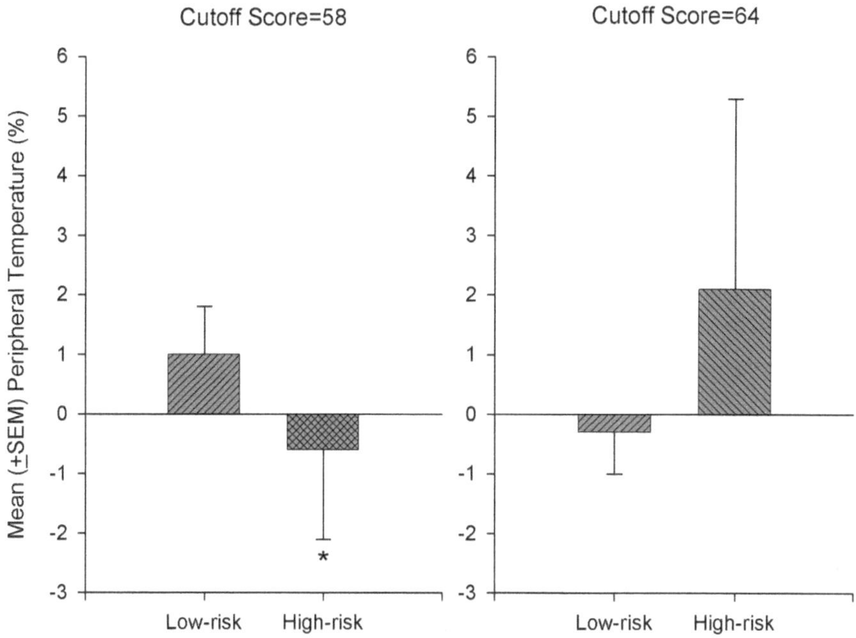

Fig. 7.3 Mean (± SEM) TEMP (%) in high- and low-risk Internet abusers, with cutoff scores of 58 or 64 in the CIAS screening of Internet addiction. *$p < 0.05$, versus low-risk group with the same cutoff score

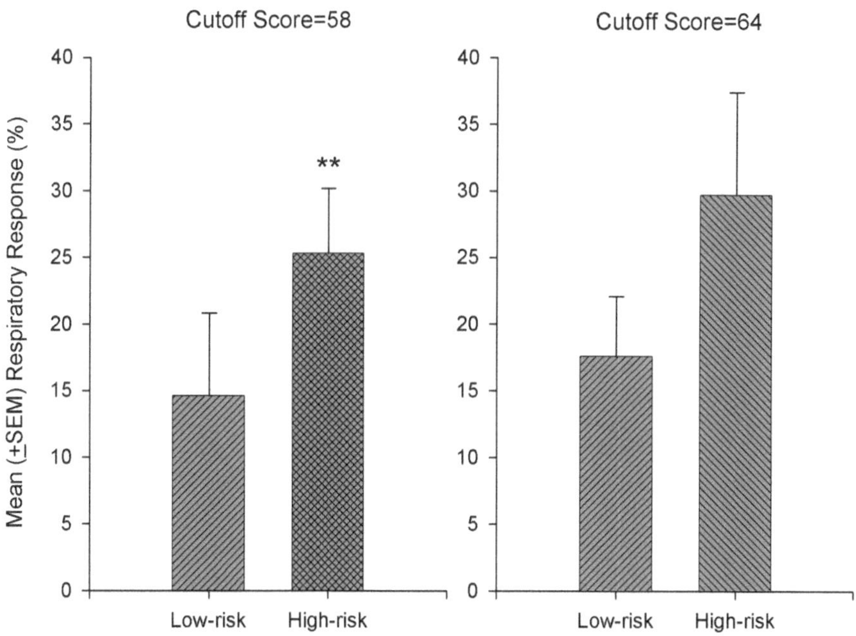

Fig. 7.4 Mean (± SEM) RESPR (%) in high- and low-risk Internet abusers, with cutoff scores of 58 or 64 in the CIAS screening of Internet addiction. **$p < 0.01$, versus low-risk group with the same cutoff score

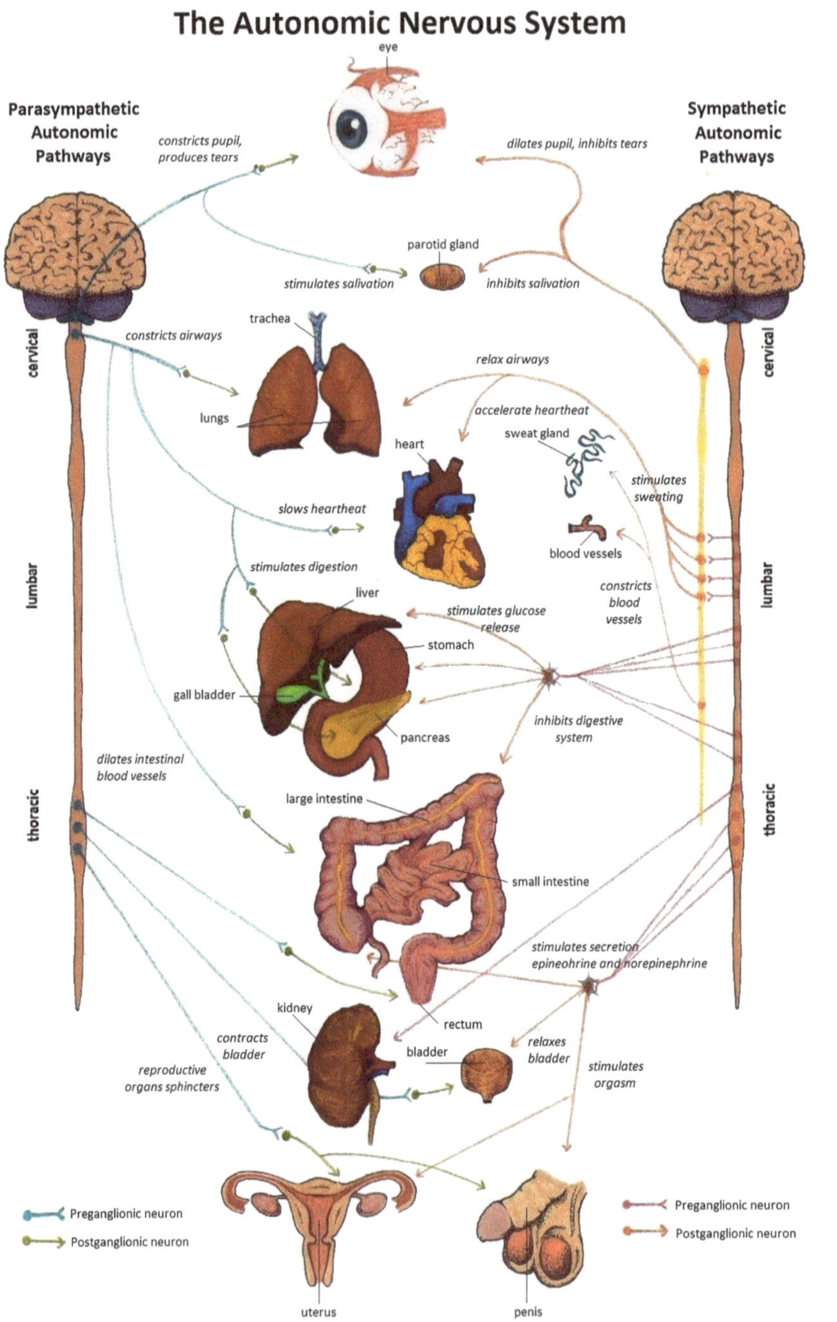

Fig. 7.5 The autonomic nervous system. The schematic figure depicts the target organs and functions served by the sympathetic and parasympathetic fibers of the autonomic nervous system

activated, the parasympathetic division is antagonized, and vice versa. This is referred to as sympathetic-parasympathetic antagonism (Carlson 2007). However, our previous data seemingly contradict the notion of sympathetic-parasympathetic antagonism. One question is why the sympathetic division is activated with regard to BVP, TEMP, and RESPR, and the parasympathetic division responds with changes in SC at the same time. This issue warrants further investigations.

Similarly paradoxical findings related to sympathetic-parasympathetic antagonism have been reported in a number of studies. For example, a study of voluntary control of physiological feedback affecting autonomic activity, found that heart rate feedback training reduced heart rate, and the deceleration in heart rate was linked to an increase in SC magnitude. The results showed a contradiction between the sympathetic division (i.e., SC responses) and parasympathetic division (i.e., heart rate activity; Gatchel 1976). Another study investigated how pathological worry affects heart rate variability arising from the sympathetic-parasympathetic nervous system. Their results indicated that threat stimuli (e.g., worrisome events) inhibited heart rate variability, and this effect was likely attributable to the inhibition of parasympathetic activity and simultaneous suppression of the sympathetic response (Weng and Teng 2005). A critical review reported that the activity arising from the sympathetic-parasympathetic divisions of the autonomic nervous system cannot be understood in terms of an antagonism of sympathetic and parasympathetic nervous activities on the level of each single internal gland or bodily organ, because some glands and organs are controlled by the single sympathetic or the single parasympathetic division of autonomic nervous system activity (Meyers 1959). For illustrative purposes; sympathetic but not parasympathetic nerves innervate nictitating membranes, most blood vessels, and sweat glands. On the other hand, the lacrimal gland, ciliary body, and iridal muscle are controlled by parasympathetic nerves, but not sympathetic nerves. Vascular smooth muscles are governed by alterations of activity in parasympathetic fibers. Although salivary glands are mediated by the sympathetic-parasympathetic divisions, two different types of cells within the salivary gland receive either sympathetic or parasympathetic innervations. In conclusion, internal glands and bodily organs may synergize but not antagonize each other through the sympathetic-parasympathetic divisions (Meyers 1959). Therefore, our previous contradictory data showing sympathetic and parasympathetic activity with regard to BV, TEMP, RESPR, and SC, are consistent with the idea that sympathetic and parasympathetic nervous system do not necessarily antagonize each other.

7.2.2 *Explanations for Contradictory Data in Four Psychophysiological Measurements in Internet Addiction: Rewarding and Aversive Properties of Addiction*

Our contradictory findings indicate that the observed differences in BV, TEMP, and RESPR in Internet addicts may be attributable to reactivation of the sympathetic division, and that alterations in SC in Internet addicts may result from parasympathetic activity. The hypothesis of the rewarding and aversive properties of addiction may help to explain this contradictory data.

According to this hypothesis, addictive stimuli and their associated cues result in both rewarding and aversive effects, which influence the sympathetic-parasympathetic divisions of autonomic nervous system activity. Two lines of research have elucidated the activity of the sympathetic-parasympathetic divisions of the autonomic nervous system in humans and animals when individuals are confronted with rewarding and aversive events (Bradley et al. 2008; Ettenberg and McFarland 2003; Firestone and Douglas 1975; Inagaki et al. 2005). For example, a recent animal study investigated the relationship between reward conditioning and activity of the autonomic nervous system. This study showed that both conditioned stimuli (i.e., cue) and unconditioned stimuli (i.e., reward) significantly increased heart rates (Inagaki et al. 2005). Additionally, a previous animal study examined the effect of the dopamine D2 receptor antagonist haloperidol on heroin consumption-related automatic activity and behavioral performance (in terms of reward and motivation). This study found that haloperidol attenuated heart rate and running speed following heroin injections. These results demonstrate that the blocking of the dopamine D2 receptor decreased the rewarding effect of heroin and reduced heart rate (Ettenberg and McFarland 2003). A prior human study assessed SC and heart rate under reward, punishment, and reward + punishment conditions. The results indicated that SC was not significantly different between these three groups, but the heart rate response was stronger in the reward condition than in the other two conditions (Firestone and Douglas 1975). Another human study examined how arousal in response to pleasant, unpleasant, and neutral pictures influenced autonomic activation. The subjects' pupil diameters and SC responses were increased when encountering pleasant and unpleasant pictures, but heart rate was decreased in relation to the occurrence of the unpleasant picture (Bradley et al. 2008).

On the other hand, the way in which aversive stimuli induce sympathetic-parasympathetic activity of the autonomic nervous system remains unclear (Campbell and Ampuero 1985; Ditto et al. 1987; Marsh et al. 2008; Miller and Ditto 1988, 1989, 1991; Olafsdottir et al. 2001; Palomba et al. 2000; Sirota and Schwartz 1976; Weng and Teng 2005). One study tested cardiac performance while watching an unpleasant film and showed that sympathetic nervous system activity was enhanced when subjects encountered the unpleasant film, including an increase in heart rate and electrodermal activation (Palomba et al. 2000). A facial expression study demonstrated that sad facial expressions (i.e., aversive stimuli) facilitated parasympathetic activity and reduced sympathetic responses, with lower SC (Marsh et al. 2008). An animal study examined how conditioned heart rate was activated in response to aversive classical conditioning with conditioned stimuli (e.g., light or tone) and unconditioned stimuli (e.g., electric shock). The authors reported that the presentation of the conditioned stimulus, which was associated with an aversive unconditioned stimulus, produced a prolonged deceleration in heart rate (Campbell and Ampuero 1985). The team of Miller and Ditto has found that subjects who were exposed to an aversive stimulus in a video-game avoidance task exhibited an increase in heart rate and peripheral vascular responses

(Miller and Ditto 1988, 1989). Moreover, they have also shown that aversive psychological stress could elicit strong sympathetic activity, such as an increase in cardiovascular responses (Miller and Ditto 1991). A previous human study trained female subjects to voluntarily control their heart rate under aversive conditions using a biofeedback device, and reported that all of the subjects could voluntarily decelerate their heart rate to relieve anxiety and fear reactions (Sirota and Schwartz 1976). Their evidence suggests that aversive stimuli increase the activity of the sympathetic nervous system through enhanced heart rate.

An old hypothesis developed by Silvestrini (1990) to explain how aversive stress activates the autonomic nervous system suggests that a stress-related stimulus may elicit both orthodox and paradoxical stress reactions. These two stress reactions are reflected in the response of the sympathetic system when individuals encounter emergency situations. The orthodox responses consist of pupil dilatation, increased blood pressure or heart rate, the concentration of blood flow at specific organs (e.g., the heart and muscles), the facilitation of blood coagulation, an increase in the metabolism of glycogen to glucose, the inhibition of some instinctual drives (e.g., hunger and sex), increases in alertness, mental activity, and muscle strength. These orthodox responses result in the effects of analgesia in physical and mental conditions. On the other hand, paradoxical stress responses are thought to be a common basis for some abnormal conditions including depression, panic attack, obesity, sexual deviation, alcoholism, and drug addiction. The paradoxical stress response produces an effect opposite to the orthodox response. For example, the orthodox stress response is speculated to decrease hunger and sex, but the paradoxical stress response excessively increases hunger and sex drives, eventually resulting in obesity and impotence. Silverstrini explained the paradoxical stress responses induced by some abnormal conditions and said, "On the basis of my empirical clinical experience, I have proposed that depression is associated with a state in which stress produces mental pain, rather than analgesia" (Silvestrini 1990, p. 7). Internet addiction also involves these abnormal and psychopathological domains (e.g., depression, obesity, alcoholism, and drug addiction) and therefore is expected to induce a paradoxical stress response, resulting in the effect of mental pain but not analgesia.

Based on the prior data, we suggest that a rewarding event (or the anticipation) may trigger sympathetic nervous system activity (e.g., an increase in heart rate and SC), whereas an aversive event may elicit parasympathetic nervous system activity (e.g., a decrease in heart rate and SC). Nevertheless, it is of importance to remember some of the aforementioned inconsistencies of sympathetic-parasympathetic divisions of the autonomic nervous system, in the context of rewarding and aversive events.

In summary, individuals with Internet addiction are suggested to expose to these two contradictory rewarding and aversive processes. When surfing the Internet, Internet abusers may encounter a rewarding process. However, when they do not use the Internet, Internet abusers may experience aversive feelings that may be related to withdrawal symptoms (Young 1996). The autonomic activity hypothesis of Internet addiction is proposed in this regard; suggesting that the sympathetic nervous system is activated by the rewarding properties of Internet addiction. However, we found that SC did not follow this viewpoint because

Internet abusers exhibited lower SC than normal subjects. Therefore, the SC response is likely related to parasympathetic activity.

7.2.3 Drug Addiction Reflected in the Autonomic Nervous System: Comparison with Internet Addiction

In contrast to Internet addiction, most addictive drugs exert their effects in the sympathetic-parasympathetic divisions of the autonomic nervous system with apparent consistency, indicating that the sympathetic nervous system, but not parasympathetic activity, is more active in drug addiction (Table 7.3; Brunelle et al. 2006; Fishbein et al. 2005; Henry et al. 2012). For example, some addiction studies related to betel chewing found that consumption of a small amount of a betel nut can induce more cardiovascular responses, suggesting that it results from sympathetic nervous activity, whereas large amounts of consumption cause hyperactivity of the parasympathetic nervous system (Chu 1995, 2001, 2002). Abusers with methamphetamine

Table 7.3 Comparisons of heart rate/blood volume pulse, skin conductance, finger temperature, and respiratory rate of the autonomic nervous system for Internet addiction and drug addiction

Autonomic nervous system activity					
Type	Characteristic				Note
	HR/BVP	SC	TEMP	RESPR	
Internet addiction	+(S)	−(Para)	−(S)	+(S)	
Betel chewing	+(S)	+(S)	N/A	N/A	Small amount →Sympathetic activity (↑) Large amount →Parasympathetic activity (↑)
Methamphetamine	+(S)	N/A	N/A	N/A	
Stimulant user (alcohol challenge)	+(S)	N/A	N/A	N/A	
Marijuana user	None	N/A	N/A	N/A	
Hallucinogen user	None	N/A	N/A	N/A	
Polydrug abuse (cocaine, heroin, alcohol, marijuana, and methamphetamine)	N/A	+(S)	N/A	N/A	
Alcohol (withdrawal state)	+(S)	N/A	N/A	+(S)	
Alcohol user	+(S)	N/A	−(S)	N/A	
Opiate user (heroin, morphine)	−(Para)	N/A	N/A	−(Para)	
Naltrexone therapy (+) or Methadone therapy (−)	+(S)	N/A	N/A	N/A	

HR heart rate; *BVP* blood volume pulse; *TEMP* peripheral temperature; *RESPR* respiratory response; *S* sympathetic nervous system; *Para* parasympathetic nervous system; *N/A* not applicable

dependence have been shown to have decreased heart rate variability, reduced parasympathetic activity, and attenuated heartbeat variability, while they also show increased sympathetic activity with heart rate facilitation (Henry et al. 2012). A comparative investigation of drug addiction among psychostimulant, marijuana, and hallucinogen abusers that assessed the activity of the autonomic nervous system found that psychostimulants abusers had significantly higher ethanol-induced heart rate increases than non-abusers. However, alcohol-induced heart rate in marijuana abusers and hallucinogen abusers was not significantly different compared to controls. This suggests that psychostimulant addiction can elicit hyperactivity of the sympathetic nervous system (Brunelle et al. 2006). A recent study of polydrug abuse, including cocaine, heroin, alcohol, marijuana, and amphetamine, indicated that drug abusers had a stronger SC response and worse performance on Gambling and Rogers Decision Making Tasks, suggestive of hyperactivity of the sympathetic nervous system in drug addiction (Fishbein et al. 2005). A review article related to alcohol addiction suggested that animals with alcohol withdrawal symptoms had increased heart rate and RESPR (Becker 2000). However, acute and chronic alcohol administration has been shown to enhance heart rate and inhibit TEMP (Boschloo et al. 2011; Johnson et al. 1986), indicating sympathetic nervous system activity, regardless of withdrawal and administration phases.

Opiate addiction exerts effects on the sympathetic-parasympathetic divisions of the autonomic nervous system through different means. Some studies of opiate abusers, such as heroin and morphine, demonstrate that dogs that receive chronic injections of morphine exhibit decreased heart rate and RESPR, indicative of parasympathetic nervous system activity (Napier et al. 1998). The opioid receptor antagonist naltrexone has been used for opiate detoxification and was shown to have an inverse effect, in which subjects had increased heart rate (Hoffman et al. 1998). Methadone, an opioid receptor agonist, activated sympathetic activity, with heart rate acceleration (Chang et al. 2012; Huang et al. 2012). Altogether, opiate abusers exhibit activation of the parasympathetic system rather than the sympathetic system. This is very different from other drug addictions (Table 7.3).

Internet addiction studies have reported contradictory data, in which the sympathetic nervous system is more highly active, with increases in heart rate and RESPR and a decrease in TEMP. Simultaneously, the parasympathetic nervous system is active, with a decrease in SC. These autonomic activity data are slightly different from those on drug addiction.

7.3 The Connections: Autonomic Nervous System and the Brain

To our knowledge, few studies have investigated how neural activity in the brain connects and governs the activity of the sympathetic-parasympathetic divisions of the autonomic nervous system (Hosoya et al. 1991; Leone et al. 2006; McAllen and May 1996; Montenegro et al. 2011; Okano et al. 2013). For example, a recent study used transcranial direct-current stimulation to assess the relationship between the temporal and insular cortices and autonomic nervous system, reflected by exercise performance. These authors suggested that stimulation of the temporal and insular cortices by transcranial direct-current stimulation modulated the activity of the autonomic nervous system and the perception of exertion and performance during maximal exercise (Okano et al. 2013). Moreover, another study showed that the application of transcranial direct-current stimulation in the left temporal lobe in athletes modulated heart rate variability. Additionally, it increased parasympathetic activity but decreased sympathetic activity and sympatho-vagal balance (Montenegro et al. 2011). Furthermore, an electrical recording study in an animal model found that the activity of medullary premotor neurons in the brainstem was correlated with the contraction of cutaneous vasoconstrictor postganglionic fibers. This suggests that brainstem neurons govern postganglionic sympathetic nerves (McAllen and May 1996). A previous review of neuroimaging and pain in humans reported that many of the brain areas related to pain functions directly mediate the activity of the sympathetic-parasympathetic divisions of the autonomic nervous system. A neuroanatomical study showed that the brain pain system is highly associated with the autonomic nervous system, and lamina 1 neurons in the brain receive pain and visceral information from relevant visceral organs that is then transmitted to the spinothalamocortical pathway in central brain structures. Interoceptive visceral messages project to the viscerosensory cortex, including the right anterior insula and orbitofrontal cortices and nociception structures, such as the anterior cingulate cortex (Leone et al. 2006). Alternatively, an afferent axon tracing study investigated the neural pathways between the hypothalamic paraventricular nucleus and peripheral sympathetic preganglionic neurons. The results indicated that the hypothalamic paraventricular nucleus projects descending inputs to sympathetic preganglionic neurons, indicated by the labeling of the anterograde transport of *Phaseolus vulgaris* leukoagglutinin (i.e., a protein derived from kidney beans and used as an anterograde tracer). Sympathetic preganglionic neurons retrogradely project ascending fibers to the hypothalamic paraventricular nucleus, revealed by cholera toxin subunit B (i.e., a retrograde tracer). Therefore, the hypothalamus paraventricular nucleus interacts with peripheral sympathetic preganglionic neurons in the autonomic nervous system (Hosoya et al. 1991).

In summary, some neural substrates of the brain indeed connect to the autonomic nervous system. The hypothalamic paraventricular nucleus reciprocally projects to sympathetic preganglionic neurons. Medullary premotor neurons in the brainstem might control the activity of postganglionic sympathetic nerves. Numerous neural substrates of pain in the brain, including the right anterior insula, orbitofrontal cortex, and anterior cingulate cortex, govern autonomic nervous activity. The temporal and insular cortices can modulate autonomic nervous system cardiac activity and the perception of exertion and performance during strenuous exercise. How central brain substrates control the peripheral sympathetic-parasympathetic divisions of the autonomic nervous system remains uncertain and needs to be scrutinized in future studies.

7.4 Internet Addiction and the Neural Circuits of Brain Reward

The neural circuitry of reward has long been demonstrated in animal models, indicating the existence of dopamine reward systems, including the mesolimbic and mesocortical dopamine systems (Di Chiara 1998; Wise 1988, 2008; Wise and Rompre 1989). The mesolimbic dopamine system is defined as the neural projection from the ventral tegmental area to many brain areas of the limbic system, such as the amygdala, nucleus accumbens, hippocampus, and olfactory cortex. The mesocortical dopamine system projects from the ventral tegmental area to the prefrontal cortex (Oades and Halliday 1987). The function of the mesolimbic dopamine system is proposed as the governance of rewarding and hedonic effects (Koob and Swerdlow 1988). However, the mesocortical dopamine system plays a crucial role in high-level cognitive functions, such as logical thinking, reasoning, and planning, although some reports have shown that its function may also incorporate rewarding and hedonic effects, similar to the mesolimbic dopamine system (Fibiger and Phillips 1988). To review the literature of Internet addiction related to neural circuitry in the brain, evidence of Internet addiction in humans has been inconsistent with drug addiction data in animals (Dong et al. 2012a, b; Han et al. 2010; Ko et al. 2009; Liu et al. 2010; Lin et al. 2012; Lorenz et al. 2013; Sun et al. 2012; Weng et al. 2013). For example, a diffusion tensor imaging study indicated that subjects with Internet gaming addiction showed greater fractional anisotropy (i.e., a value that is calculated from a marker of diffusion along the axon and diffusion perpendicular to the axon) due to an increase in white matter integrity in the thalamus and left posterior cingulate cortex compared with normal subjects. Moreover, greater volume of white matter in the thalamus was associated with the severity of Internet addiction (Dong et al. 2012a). A similar study that used a diffusion tensor imaging approach related to Internet addiction showed that subjects with Internet addiction had significantly decreased fractional anisotropy in orbitofrontal white matter, corpus callosum, cingulum, inferior fronto-occipital fasciculus, and corona radiation internal and external capsules (Lin et al. 2012). A recent study of online gaming addiction suggested that abnormal gray matter and white matter volumes may be related to online gaming abuse. These abnormal brain areas included the right orbitofrontal cortex, insular cortex, and right supplementary motor cortex. Furthermore, white matter volume was decreased in the right genus of the corpus callosum, frontal cortex, and right external capsule, whereas gray matter volume increases were correlated with the right orbitofrontal cortex, insular cortex, and right external capsule in Internet gaming abusers (Weng et al. 2013). A recent functional magnetic resonance imaging (fMRI) study of Internet addiction to the online game World of Warcraft provided supporting evidence implicating similar brain regions, suggesting that some common neural substrates may mediate online game craving behaviors in Internet addiction (Ko et al. 2009). The common neural substrates of the brain comprise the right orbitofrontal cortex and bilateral anterior cingulate cortex. Moreover, the right nucleus accumbens, medial frontal cortex, dorsolateral prefrontal cortex, and right caudate nucleus may also be involved in craving behavior in Internet addiction (Ko et al. 2009). Thus, reward-related brain areas may mediate the specific form of Internet addiction, such as the online game Word of Warcraft; however, another study with the online game, Counter Strike, has not implicated neural reward mechanisms in this category of Internet gaming addiction (Montag et al. 2012). This raises questions as to whether neural reward pathways are selectively implicated in internet (gaming) addiction dependent on game-specific characteristics (e.g., role-player, point of view, the accumulation of "point", story narrative etc.).

With regard to craving behavior and cue-induced Internet addiction, a brain mapping study of Internet video game play showed that the anterior cingulate cortex and orbitofrontal cortex were significantly activated when subjects encountered Internet video game cues (Han et al. 2010). Another study found greater brain activity in a cue/picture exposure condition, including in the dorsolateral prefrontal cortex, bilateral temporal lobe, cerebellum, right inferior parietal lobe, right cuneus, right hippocampus, parahippocampus, parahippocampal gyrus, and left caudate nucleus (Sun et al. 2012). These results suggest that these brain areas may be involved in cue-induced Internet addiction or craving behavior.

A recent study that used resting-state fMRI brain mapping found that Internet gaming addicts exhibited higher brain activity in the brainstem, inferior parietal lobule, left posterior cerebellum, and left middle frontal gyrus. Additionally, regional homogeneity measures of the temporal, occipital, and parietal lobes were significantly decreased in Internet gaming abusers. The authors suggest that these brain areas may be involved in sensory and motor coordination or visual and auditory function, but the data did not suggest the involvement of these areas in reward and hedonic effects (Dong et al. 2012b). Another study indicated that Internet gaming abusers had significantly greater increases in regional homogeneity in the cerebellum, brainstem, right cingulate gyrus, bilateral parahippocampus, right frontal lobe, left superior frontal gyrus, left precuneus, right postcentral gyrus, right middle occipital gyrus, right inferior temporal gyrus, left superior temporal gyrus, and middle temporal gyrus. The results suggested that the neural networks among the cerebellum, limbic cortex, brainstem, and frontal cortex might be correlated with the rewarding and hedonic effects of Internet addiction (Liu et al. 2010).

In summary, the results of human Internet addiction studies are more complex than the findings from animal drug addiction research. The mesolimbic and mesocortical dopamine systems may comprise the framework for Internet addiction. Unfortunately, the dopamine hypothesis, which was developed using evidence from animal studies, is not fully consistent with Internet addiction findings related to the neural circuitry of the brain. This suggests a need for the development of a novel hypothesis of Internet addiction in the future.

7.5 Internet Addiction: Pathogenesis and Interventions

A growing body of evidence has shown that Internet addiction can induce depression or that the pathogenesis of Internet addiction is induced by uncontrolled Internet use; moreover, Internet abusers often show numerous psychopathological symptoms and behaviors related to Internet addiction (Karim and Chaudhri 2012; Carli et al. 2013). For example, Internet abusers can exhibit uncontrollable use of the Internet, distressing feelings, and time-consuming habits that result in social and occupational difficulties (Shapira et al. 2000). Some Internet abusers suffer from depression and guilty feelings. They may also exhibit an increase in the probability of aggressive behavior when exposed to long-term use of the Internet (Treuer et al. 2001). Therefore, many Internet abusers state that they encounter relational, academic, familial, and occupational impairments (Young 2007).

Internet addition was not defined as a mental disorder by the *Diagnostic and Statistical Manual of Mental Disorders*, 4th edition (DSM-IV), but the current version, the DSM-5, includes Internet gaming addiction (American Psychiatric Association 2013). Internet addiction was first proposed as a psychological or psychopathological disorder by Ivan Goldberg in 1995 (Eppright et al. 1999; Garrison and Long 1995). He adapted the psychopathological symptoms of pathological gambling to describe the compulsive symptoms of Internet addiction, such as pathological Internet use and uncontrolled Internet use that results in physical, social, psychological, and occupational dysfunction (Young 1998). Moreover, the characteristics of Internet addiction may be associated with symptoms of substance dependence that involve mood alterations, tolerance, abstinence, and withdrawal symptoms (Griffiths 1995). Internet addiction is also correlated with some psychopathological behaviors, such as depression, anxiety, hostility, psychoticism, interpersonal sensitivity, attention-deficit/hyperactivity disorder, obsessive-compulsive disorder, novelty seeking, and social anxiety disorder (Carli et al. 2013; Cho et al. 2013). However, all of these psychopathological behaviors are not required simultaneously to define an Internet abuser. Rather, some number of characteristics may present in one Internet addict, while additional and/or different characteristics may typify another Internet addict. Similar to many psychological disorders, e.g. substance addiction and depression, the presentation of symptoms can be heterogenous across patients. The symptoms and psychopathological behaviors associated with Internet addiction are very diverse. Thus, therapeutic interventions for Internet addiction may be complicated.

Recently, some studies have examined the relationship between Internet addiction and psychopathology. For example, Cho et al. (2013) found that withdrawal and anxiety/depression symptoms during childhood could predict the occurrence of Internet addiction. These authors suggested that clinicians should consider withdrawal and anxiety/depression behaviors during childhood to protect humans with an increased vulnerability for Internet addiction (Cho et al. 2013). The psychopathological compulsive behavior associated with Internet addiction is similar to that in alcohol use, resulting from implicit cognition to the exclusion of explicit cognition to control substance abuse (Yen et al. 2011). Xiuqin et al. (2010) investigated pathological symptoms, personality, and parental rearing styles in adolescents with Internet addiction, suggesting that Internet abusers might have higher obsessive-compulsive behavior, interpersonal sensitivity, depression, anxiety, hostility, and paranoid ideation. Additionally, the personality traits of Internet abusers trend toward a lower magnitude of extraversion and a higher degree of psychoticism, while parental rearing styles often lean toward being over-intrusive, punitive, and lacking in responses during parent-child interactions (Xiuqin et al. 2010). A recent study of Internet addiction suggested that obsessive-compulsive symptoms develop before Internet abusers become addicted to the Internet. Depression, anxiety, hostility, interpersonal sensitivity, and psychoticism are present after Internet addiction develops (Dong et al. 2011). Our recent study further showed that Internet addiction is associated with depressive states but not depressive traits (Huang et al. 2013). This suggests that the psychopathological depression associated with Internet addiction is not related to a depressive personality trait. Instead, psychopathological depressive symptoms are a temporary depressive state. Therefore, interventions for psychopathological depression should target temporary depressive states rather than permanent depressive traits. Interestingly, the notion that depressive state is more important than personality trait as a vulnerability factors for depression has been challenged by recent findings (Mehroof and Griffiths 2010; Tsai et al. 2009). This requires further scrutiny in future studies.

In conclusion, some crucial points are the following. Obsessive-compulsive symptoms disorder, depression, anxiety, hostility, and interpersonal sensitivity may be common psychopathological symptoms of Internet addiction. Obsessive-compulsive behavior can also effectively predict the occurrence of Internet addiction. Moreover, obsessive-compulsive behavior is a critical precursor for Internet addiction. Depression, anxiety, hostility, and interpersonal sensitivity are psychopathological symptoms of Internet addiction; thus, these symptoms and behaviors are the sequelae that appear after Internet addiction develops. Dissociations between depressive states and depressive traits with regard to psychopathological depressive behavior are important considerations for interventions. Finally, the occurrence of Internet addiction may involve many factors in addition to psychopathological symptoms, including personality traits and parental rearing styles.

7.6 Conclusions

Interventions for Internet addiction should include pharmacological and non-pharmacological approaches. The pathogenesis of Internet addiction involves complicated and multiple psychopathological and behavioral symptoms. The role of the autonomic nervous system and the neural circuitry involved in the multiple psychopathological symptoms of Internet addiction require elucidation to enable development of novel and effective interventions. The development of new drugs that focus on the sympathetic-parasympathetic divisions of the peripheral autonomic nervous system is also a crucial line of research. With regard to interventions associated with the peripheral autonomic nervous system, the antagonism of sympathetic nervous system activity may be a key way to prevent against Internet addiction. Thus, research in this basic area requires replication and verification in the future studies.

Acknowledgments This work was supported by grants NSC 101-2413-H-431-007 and NSC 102-2410-H-431-005-MY3 from the National Research Council of the Republic of China. I would like to thank my students, Mr. Cai N Cheng and Mr. Alan Bo Han He, for drawing Fig. 7.5.

References

American Psychiatric Association (2013) Diagnostic and statistical manual of mental disorders, 5th edn. American Psychiatric Association, Washington, DC

Becker HC (2000) Animal models of alcohol withdrawal. Alcohol Res Health 24:105–113

Boschloo L, Vogelzangs N, Licht CM et al (2011) Heavy alcohol use, rather than alcohol dependence, is associated with dysregulation of the hypothalamic-pituitary-adrenal axis and the autonomic nervous system. Drug Alcohol Depend 116:170–176

Bradley MM, Miccoli L, Escrig MA, Lang PJ (2008) The pupil as a measure of emotional arousal and autonomic activation. Psychophysiology 45:602–607

Brunelle C, Barrett SP, Pihl RO (2006) Psychostimulant users are sensitive to the stimulant properties of alcohol as indexed by alcohol-induced cardiac reactivity. Psychol Addict Behav 20:478–483

Campbell BA, Ampuero MX (1985) Dissociation of autonomic and behavioral components of conditioned fear during development in the rat. Behav Neurosci 99:1089–1102

Carli V, Durkee T, Wasserman D et al (2013) The association between pathological internet use and comorbid psychopathology: a systematic review. Psychopathology 46:1–13

Carlson NR (2007) Structure of the nervous system. In: Carlson M, Carlson NR (eds) Physiology of Behavior, 9th edn. Pearson Allyn and Bacon, Boston, pp 68–101

Chang LR, Lin YH, Kuo TB et al (2012) Cardiac autonomic modulation during methadone therapy among heroin users: a pilot study. Biol Psychiatry 37:188–193

Cho SM, Sung MJ, Shin KM et al (2013) Does psychopathology in childhood predict internet addiction in male adolescents? Child Psychiatry Hum Dev 44:549–555

Chu NS (1995) Effect of betel chewing on RR interval variation. J Formos Med Assoc 94:106–110

Chu NS (2001) Effects of betel chewing on the central and autonomic nervous systems. J Biomed Sci 8:229–236

Chu NS (2002) Neurological aspects of areca and betel chewing. Addict Biol 7:111–114

Di Chiara G (1998) A motivational learning hypothesis of the role of mesolimbic dopamine in compulsive drug use. J psychopharmacol 12:54–67

Ditto B, Miller S, Maurice S (1987) Age differences in the consistency of cardiovascular response patterns in healthy women. Biol Psychol 25:23–31

Dong G, DeVito E, Huang J, Du X (2012a) Diffusion tensor imaging reveals thalamus and posterior cingulate cortex abnormalities in internet gaming addicts. J Psychiatr Res 46:1212–1216

Dong G, Huang J, Du X (2012b) Alterations in regional homogeneity of resting-state brain activity in internet gaming addicts. Behav Brain Funct 8:41

Dong G, Lu Q, Zhou H, Zhao X (2011) Precursor or sequela: pathological disorders in people with Internet addiction disorder. PLoS ONE 6:e14703

Eppright T, Allwood M, Stern B, Theiss T (1999) Internet addiction: a new type of addiction? Mo Med 96:133–136

Ettenberg A, McFarland K (2003) Effects of haloperidol on cue-induced autonomic and behavioral indices of heroin reward and motivation. Psychopharmacology 168:139–145

Fibiger HC, Phillips AG (1988) Mesocorticolimbic dopamine systems and reward. Ann N Y Acad Sci 537:206–215

Firestone P, Douglas V (1975) The effects of reward and punishment on reaction times and autonomic activity in hyperactive and normal children. J Abnorm Child Psychol 3:201–216

Fishbein D, Hyde C, Eldreth D et al (2005) Cognitive performance and autonomic reactivity in abstinent drug abusers and nonusers. Exp Clinic Psychopharmacol 13:25–40

Garrison J, Long P (1995) Getting off the superhighway. Health 9:20–22

Gatchel RJ (1976) The effect of voluntary control of heart rate deceleration on skin conductance level: an example of response fractionation. Biol Psychol 4:241–248

Griffiths MD (1995) Towards a risk factor model of fruit machine addiction: a brief note. J Gambl Stud 11:343–346

Han DH, Kim YS, Lee YS et al (2010) Changes in cue-induced, prefrontal cortex activity with video-game play. Cyberpsychol Behav Soc Netw 13:655–661
Henry BL, Minassian A, Perry W (2012) Effect of methamphetamine dependence on heart rate variability. Addict Biol 17:648–658
Hoffman WE, McDonald T, Berkowitz R (1998) Simultaneous increases in respiration and sympathetic function during opiate detoxification. J Neurosurg Anesthesiol 10:205–210
Hosoya Y, Sugiura Y, Okado N et al (1991) Descending input from the hypothalamic paraventricular nucleus to sympathetic preganglionic neurons in the rat. Exp Brain Res 85:10–20
Huang ACW, Chen HE, Wang YC, Wang LM (2013) Internet abusers associate with a depressive state but not a depressive trait. Psychiatr Clinic Neurosci 68(3):197
Huang WL, Lin YH, Kuo TB et al (2012) Methadone-mediated autonomic functioning of male patients with heroin dependence: the influence of borderline personality pattern. PLoS ONE 7:e37464
Inagaki H, Kuwahara M, Tsubone H (2005) Changes in autonomic control of heart associated with classical appetitive conditioning in rats. Jpn Assoc Lab Anim Sci 54:61–69
Johnson RH, Eisenhofer G, Lambie DG (1986) The effects of acute and chronic ingestion of ethanol on the autonomic nervous system. Drug Alcohol Depend 18:319–328
Karim R, Chaudhri P (2012) Behavioral addictions: an overview. J Psychoact Drugs 44:5–17
Ko CH, Liu GC, Hsiao S et al (2009) Brain activities associated with gaming urge of online gaming addiction. J Psychiatr Res 43:739–747
Koob GF, Swerdlow NR (1988) The functional output of the mesolimbic dopamine system. Ann N Y Acad Sci 537:216–227
Leone M, Proietti CA, Mea E et al. (2006) Neuroimaging and pain: a window on the autonomic nervous system. Neurol Sci 27: S134–S137
Lin F, Zhou Y, Du Y et al (2012) Abnormal white matter integrity in adolescents with internet addiction disorder: a tract-based spatial statistics study. PLoS ONE 7:e30253
Liu J, Gao XP, Osunde I et al (2010) Increased regional homogeneity in internet addiction disorder: a resting state functional magnetic resonance imaging study. Chin Med J 123:1904–1908
Lorenz RC, Kruger JK, Neumann B et al (2013) Cue reactivity and its inhibition in pathological computer game players. Addict Biol 18:134–146
Marsh P, Beauchaine TP, Williams B (2008) Dissociation of sad facial expressions and autonomic nervous system responding in boys with disruptive behavior disorders. Psychophysiology 45:100–110
McAllen RM, May CN (1996) Brainstem neurones and postganglionic sympathetic nerves: does correlation mean connection? Acta Neurobiol Exp 56:129–135
Mehroof M, Griffiths MD (2010) Online gaming addiction: the role of sensation seeking, self-control, neuroticism, aggression, state anxiety, and trait anxiety. Cyberpsychol Behav Soc Netw 13:313–316
Meyers FH (1959) A critique of the concept of sympathetic-parasympathetic antagonism. J Am Geriatr Soc 7:120–127
Miller SB, Ditto B (1988) Cardiovascular responses to an extended aversive video game task. Psychophysiology 25:200–208
Miller SB, Ditto B (1989) Individual differences in heart rate and peripheral vascular responses to an extended aversive task. Psychophysiology 26:506–513
Miller SB, Ditto B (1991) Exaggerated sympathetic nervous system response to extended psychological stress in offspring of hypertensives. Psychophysiology 28:103–113
Montag C, Weber B, Trautner P et al (2012) Does excessive play of violent first-person-shooter-video-games dampen brain activity in response to emotional stimuli? Biol Psychol 89:107–111
Montenegro RA, Farinatti PT, Fontes EB et al (2011) Transcranial direct current stimulation influences the cardiac autonomic nervous control. Neurosci Lett 497:32–36
Napier LD, Stanfill A, Yoshishige DA et al (1998) Autonomic control of heart rate in dogs treated chronically with morphine. Am J Physiol Heart Circ Physiol 275:H2199–H2210
Oades RD, Halliday GM (1987) Ventral tegmental (A10) system: neurobiology. 1. Anatomy and connectivity. Brain Res Rev 434:117–165

Okano AH, Fontes EB, Montenegro RA et al. (2013) Brain stimulation modulates the autonomic nervous system, rating of perceived exertion and performance during maximal exercise. Br J Sports Med in press. doi:10.1136/bjsports-2012-091658
Olafsdottir E, Ellertsen B, Berstad A, Fluge G (2001) Personality profiles and heart rate variability (vagal tone) in children with recurrent abdominal pain. Acta Paediatr 90:632–637
Palomba D, Sarlo M, Angrilli A et al (2000) Cardiac responses associated with affective processing of unpleasant film stimuli. Int J Psychophysiol 36:45–57
Shapira NA, Goldsmith TD, Keck PE et al (2000) Psychiatric features of individuals with problematic internet use. J Affect Disord 57:267–272
Silvestrini B (1990) The paradoxical stress response: a possible common basis for depression and other conditions. J Clinic Psychiatr 51:6–8
Sirota AD, Schwartz GE (1976) Voluntary control of human heart rate: effect on reaction to aversive stimulation: a replication and extension. J Abnorm Psychol 85:473–477
Sun Y, Ying H, Seetohul RM et al (2012) Brain fMRI study of crave induced by cue pictures in online game addicts (male adolescents). Behav Brain Res 233:563–576
Treuer T, Fabian Z, Furedi J (2001) Internet addiction associated with features of impulse control disorder: is it a real psychiatric disorder? J Affect Disord 66:283
Tsai HF, Cheng SH, Yeh TL et al (2009) The risk factors of Internet addiction—a survey of university freshmen. Psychiatry Res 167:294–299
Weng CB, Qian RB, Fu XM et al (2013) Gray matter and white matter abnormalities in online game addiction. Eur J Radiol 82:1308–1312
Weng CY, Teng MH (2005) Suppression of autonomic nervous system caused by worry. Chin. J. Psychol 47:364–379
Wise RA (1988) The neurobiology of craving: implications for the understanding and treatment of addiction. J Abnorm Psychol 97:118–132
Wise RA (2008) Dopamine and reward: the anhedonia hypothesis 30 years on. Neurotox Res 14:169–183
Wise RA, Rompre PP (1989) Brain dopamine and reward. Annu Rev Psychol 40:191–225
Xiuqin H, Huimin Z, Mengchen L et al (2010) Mental health, personality, and parental rearing styles of adolescents with Internet addiction disorder. Cyberpsychology, Behavior, and Social Networking 13:401–406
Yen JY, Yen CF, Chen CS et al (2011) Cue-induced positive motivational implicit response in young adults with Internet gaming addiction. Psychiatry Res 190:282–286
Young KS (1996) Psychology of computer use: XL. Addictive use of the Internet: a case that breaks the stereotype. Psychol Rep 79:899–902
Young KS (1998) Internet addiction: the emergence of a new clinical disorder. CyberPsychol Behav 1:237–244
Young KS (2007) Cognitive behavior therapy with Internet addicts: treatment outcomes and implications. CyberPsychol Behav 10:671–679

Chapter 8
A Short Summary of Neuroscientific Findings on Internet Addiction

Christian Montag, Éilish Duke and Martin Reuter

Abstract Neuroscientific approaches to the understanding of Internet addiction have broadened our knowledge on the biological basis related to the overuse of the Internet. The present chapter lends a short introduction to this area. Moreover, it integrates and summarizes the most important findings of this research field.

8.1 Background of the Present Section

The present chapter closes the section on neuroscientific findings on Internet addiction. In total six chapters of this book give the reader detailed insights into different approaches to investigating Internet addiction through use of neuroscientific techniques. While structuring the book and inviting authors to contribute, we became aware that it would be meaningful to ask internationally renowned researchers to summarize the findings on the nature of Internet addiction from distinct perspectives such as structural/functional MRI, PET, and molecular genetics, and also by investigating the peripheral autonomous activity using classic biopsychological tools such as recording electrodermal activity and heart rate. Clearly all these techniques have unique advantages and disadvantages, thus different approaches to Internet addiction illuminate different facets of this new potential psychopathological disorder. As only

C. Montag (✉)
Department of Psychology, University of Ulm, Ulm, Germany
e-mail: christian.montag@uni-ulm.de

M. Reuter
Center for Economics and Neuroscience, University of Bonn, Bonn, Germany
e-mail: martin.reuter@uni-bonn-diff.de

É. Duke
Department of Psychology, Goldsmiths, University of London, London, UK

M. Reuter
Department of Psychology, University of Bonn, Bonn, Germany

C. Montag and M. Reuter (eds.), *Internet Addiction*, Studies in Neuroscience, Psychology and Behavioral Economics, DOI 10.1007/978-3-319-07242-5_8

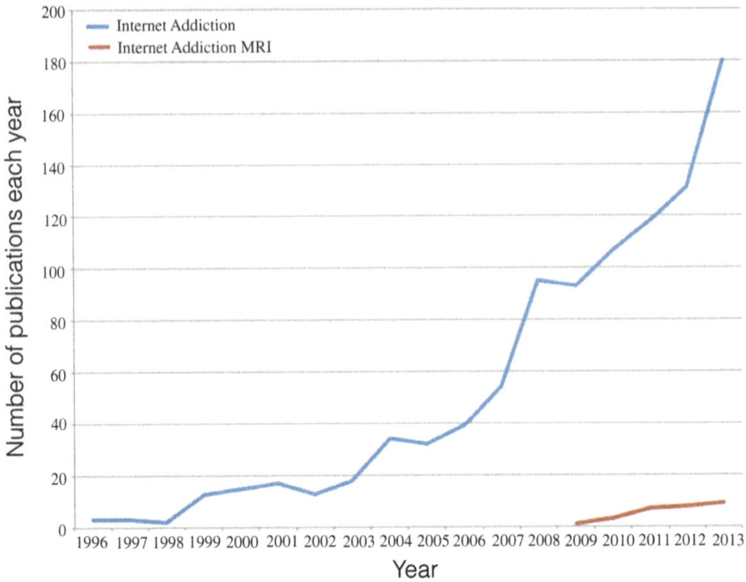

Fig. 8.1 Number of publications dealing with Internet addiction each year since 1996 split into all papers (*blue*) and those only dealing with MRI (*red*) found on pubmed.com

a small number of studies used EEG, we refrained from including a further chapter on EEG research (although this was initially planned). Nevertheless, we will briefly outline the EEG findings on Internet addiction in this synopsis. For readers not familiar with the background of the neuroscientific techniques, we refer in addition to the chapters in this book, to the extensive Appendix of the volume "Neuroeconomics", also appearing in the present series "Studies in Neuroscience, Psychology and Behavioral Economics", giving scholarly insights into a wide range of important methods.

In the following we would like to outline some general thoughts on the question of whether Internet addiction actually qualifies as a distinct psychopathological disorder. Before we aim to find a preliminary answer to this question, we first would like to underline the importance of publishing a compendium on the status quo of Internet addiction research. In this context we produced Fig. 8.1, which is discussed in the next section.

8.2 Some Numbers on Published Internet Addiction Papers

On the 24th March 2014 we searched pubmed.com for papers dealing either with the keyword "Internet addiction" or a combination of the keywords "Internet addiction" and "MRI". Without doubt MRI represents the most frequently used neuroscientific technique in the study of Internet addiction. While general research dealing with Internet addiction started in the year 1996, the first studies using MRI to gain insights into the biological underpinnings of MRI appeared 13 years later, in the year 2009. Clearly the numbers of published Internet addiction papers continue to rise dramatically each year. In the last year N = 180 papers were published on Internet addiction, totaling up to N = 967 papers since 1996. Since the year 2009 N = 28 papers have been published on Internet addiction utilizing MRI. The popularity of MRI studies in the field of neuroscientific Internet addiction research is also reflected by the inclusion of three chapters in the present book dealing with structural/functional MRI and Internet addiction. In general, the tremendous research efforts in this new field clearly mirrors the need to better understand how the Internet challenges our society and how our brains react to an increasingly complex (digital) world[1].

8.3 Cue Reactivity Paradigms, the Striatum and Internet Addiction

Although it is too early to answer the question of whether Internet addiction qualifies as a distinct behavioral addiction or otherwise, several similarities between Internet addiction and other well-known forms of addiction are striking. In the following we will summarize some of the most important overlaps.

One classic approach to the study of addiction by means of functional MRI is the use of cue reactivity experiments. Here, the drug stimulus (e.g. a computer screen with a browser) is shown to the (Internet) addict via special goggles or a mirror system while lying in the MRI scanner. By recording the brain activity in this setting, the researcher hopes to gain insights into the biological mechanisms underlying addictive tendencies. This approach to the study of addiction in humans has been adapted from 'classic' psychological research. Here, Carter and Tiffany (1999) observed in their meta-analyses that typical reaction patterns in different groups of addicts towards the exposure of their preferred drug cue can be observed in the periphery of the human body. These include elevated heart rate, higher electrodermal activity and lower skin temperature. In line with this, Huang describes in Chap. 7 in the present volume, that Internet addicts are also characterized by higher blood volume pulse and lower skin temperature, while surfing the web. Conflicting findings were observed with respect to skin conductance, which was lower in Internet addicts compared to healthy Internet users. Despite the latter finding, similarities between the bodily responses of Internet addicts compared to other types of addicts can be observed. This all naturally hints towards the idea to include cue reactivity experiments in an MRI setting, too.

---

[1] Of note, the numbers are just a rough (conservative) estimate on studies dealing with this research topic, because the inclusion of additional keywords (such as problematic Internet use) would have yielded higher results. Please see also our foreword for a short thought on the terminology.

The adoption of cue reactivity experiments in MRI paradigms suggests that the striatal brain region represents a key to understanding the psychological processes of wanting and consuming a drug, such as the Internet. Ko et al. (2009) observed stronger neuronal activity in this brain area while processing World of Warcraft stimuli in excessive gamers versus healthy controls. Sun et al. (2012) observed a similar activity in the nucleus caudatus in an identical design.[2] It is noteworthy that this kind of brain activation can also be observed in the processing of alcohol stimuli while scanning alcoholics (Vollstädt-Klein et al. 2010), smoking related stimuli while scanning smokers (Franklin et al. 2007) or heroin related stimuli while scanning heroin addicts (Li et al. 2012). Deriving from this, different kinds of addictions seem to share common biological underpinnings in the processing of drug cues. The fact that some of the aforementioned studies deal more with online video game addiction (instead of generalized Internet addiction) is mentioned as a limitation in the summary of findings below.

8.4 Dopamine, the Striatum and Internet Addiction

Functional MRI studies help to describe which brain areas play an important role in the processing of drug cues, craving or being intoxicated (e.g. by consuming too much alcohol). Besides giving these valuable insights, functional MRI clearly also possesses limitations, such as having a relatively low temporal resolution. In contrast, EEG is able to record brain processes within a more accurate time window, but is unable to sufficiently image subcortical processes. Both fMRI and EEG are unable to give direct insights into the active transmitter systems of a given brain area when recording brain activity. To close this gap, other approaches from brain imaging, such as PET or molecular genetics, need to be taken into account for a better understanding of the biochemical processes underlying Internet addiction.

It is common knowledge that the striatum, in particular the nucleus accumbens, as central brain region for the processing of and craving for rewards, is strongly innervated by the neurotransmitter dopamine, which represents a classic target to understand addiction and reward (Berke and Hyman 2000; Wise and Rompré 1989). Thereby it might not come as a surprise that one of the previous chapters dealing with the imaging technique PET revealed strong evidence for a link between dopamine and Internet addiction. Park and Kim observe in Chap. 4 that Internet addicts are characterized by a lower D_2 receptor density (something which is typical for additive tendencies in general; e.g. Blum et al. 1996). This is also in line with one of the earliest PET findings in this research field, which showed that the dopaminergic neurotransmission is triggered while playing a video game (Koepp et al. 1998).

To explain these noteworthy dopamine findings a bit more, the role of dopamine in addiction needs to be better defined. A seminal review by Volkow et al.

[2] The above mentioned studies by Ko et al. (2009) and Sun et al. (2012) reported a wide range of further brain areas of interest, which are not discussed at this point for the sake of brevity.

(2003) summarized that strong dopaminergic bursts accompany the consumption of or craving for a drug. The over-activity of the dopaminergic system when being confronted with or anticipating a drug makes it hard for the addict to fight against the actual consumption, and to focus instead on potential negative long-term consequences of the drug abuse. Here, we will also need to discuss the findings on Internet addiction and the anterior cingulate cortex (ACC), reflecting top–down processes to control the dopaminergic bursts in subcortical areas when being confronted with a drug. This is done in the next section.

Coming back to the observation that lower D_2 receptor density can be observed in Internet addicts; it is of importance to note that a lower number of these receptors could result in a higher threshold required for drug stimulation, thereby increasing the amount of drug intake necessary to achieve a similar threshold. Thanos et al. (2001) revealed that the artificial up-regulation of D_2 receptors in rats led to less alcohol consumption, whereas the down-regulation of these receptors was associated with even higher consumption of alcohol. In humans (as in other mammals), the D_2 receptor density of the striatum itself is in parts determined by genetics disposing a person more or less to become addicted. As summarized in our chapter on molecular genetics and Internet addiction (see Chap. 6), the first evidence has been provided that Internet addicts more often carry the A1+ variant of the DRD2/ANKK1 Taq Ia polymorphism (Han et al. 2007), which is linked to a 30–40 % reduction of the D_2 receptor density in striatal regions (e.g. review by Noble 2000). Moreover, the A1+ variant also seems to be associated with a lower binding capacity of dopamine at the D_2 receptor (Thompson et al. 1997). As mentioned above, carrying the A1+ variant reflects not a tendency to become Internet addicted per se but rather represents a general vulnerability factor for addictive behavior, as the A1+ variant has also been associated with a wide range of addictions including alcoholism (Munafo et al. 2007), heroin addiction (Teh et al. 2012) and smoking (De Ruyck et al. 2010).

8.5 Conflict Monitoring, Anterior Cingulate Cortex and Internet Addiction

Another key region in addiction research is the ACC. This brain area is crucially involved in conflict monitoring (Kerns et al. 2004). Conflicting situations or dilemmas play an important role in addiction research, because drug addicts are usually confronted with the lure of immediate drug consumption/reward and thereby taking into account long term negative consequences such as bad health or social repercussions.

In the context of addiction research, it has been demonstrated that, for example, opiate addicts show an attenuated error-related activity of the rostral ACC, when performing a Go/NoGo task in an fMRI setting (Forman et al. 2004). In a Go/NoGo task, participants of a study need to suppress a learned dominant behavior, such as pushing a button, when not appropriate. The poorer performance of addicts compared to healthy participants on such a task, supports the idea that monitoring functions of the ACC seem not to work properly in addicts (Lubman et al. 2004).

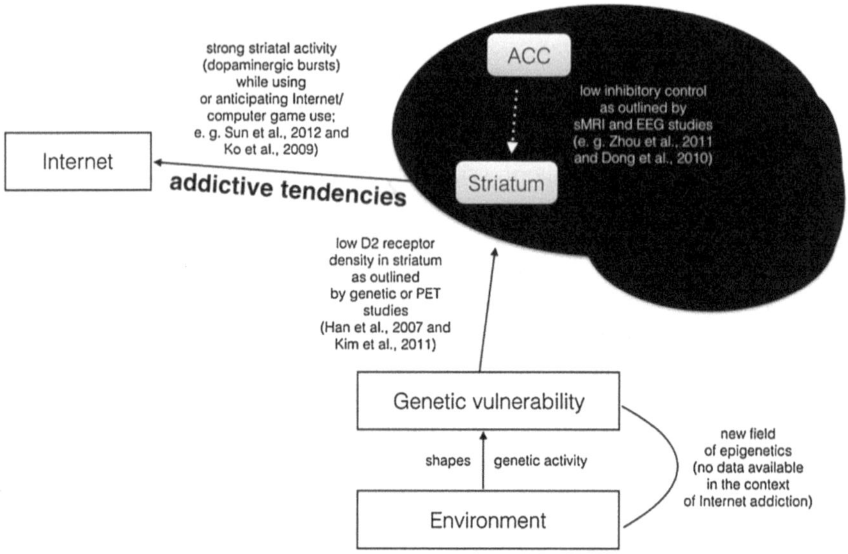

Fig. 8.2 A simplified psychobiological model of Internet addiction including key findings from Internet addiction research

This also potentially explains why addicts are prone to weigh the short-term rewards higher than the long-term negative consequences of drug misuse. In addition, cocaine addicts have been associated with lower gray matter volume of the cingulate area (Franklin et al. 2002). Although studies bridging structural with functional brain imaging are scarce, it is imaginable that a reduction of brain volume in a given brain area such as the ACC goes along with a higher chance of dysfunctionality.

The findings on the ACC-addiction link can also be transferred in part to Internet addiction research. In line with the study by Franklin et al. (2002), Internet addicts have been associated with lower gray matter density of the ACC (Zhou et al. 2011; see also Chap. 2). Moreover, a recent EEG study by Dong et al. (2010) provides further data for a dysfunction of the ACC in Internet addiction, because Internet addicts "had lower activation in the conflict detection stage than the normal group" in their Go/NoGo experiment (p. 138). A follow-up study by Dong et al. (2011a) extended these findings in the same direction using a color word Stroop experiment. The group of Internet addicts was associated with a reduced medial frontal negativity of the EEG signal going along with more errors made in the incongruent condition[3] of the experiment and longer reaction times

[3] In the color word Stroop experiment, words describing colors are presented in two conditions. The congruent condition e.g. consists of the word blue presented in blue color, whereas in the incongruent condition the word blue is presented in red color. Participants are required to name the color of the shown words (i.e. not the word printed).

compared to healthy controls. Finally Dong et al. (2011b) showed that Internet addicts respond with lower activity in the ACC to losses in a guessing task.

Taken together, these findings point towards an impairment among Internet addicts in resolving conflicts and problems in following a long term health strategy. The low inhibitory control from the prefrontal cortex, including the ACC, over the strong dopaminergic bursts from subcortical areas when being confronted with a drug, seem to represent an essential part of the biology of Internet addiction. This simplified model is also depicted in Fig. 8.2.

8.6 Summary of the Present Overview

These two examples focusing on the striatum and the anterior cingulate demonstrate that indeed several findings from Internet addiction research integrate well into findings from general addiction research. Nevertheless, it is too early to state that Internet addiction represents a distinct behavioral addiction.

First of all, an own study recently outlined that it is problematic to compare all kinds of specific and generalized Internet addictions, because the overlap in terms of co-morbidity between these different forms of Internet addiction is not large (Montag et al. 2014). Besides this, longitudinal studies describing the effects of excessive Internet usage on the brain are largely missing. In contrast to smoking, alcohol, cocaine or heroin, no substance is directly inserted into the human body. Therefore, brain changes due to Internet addiction may ultimately prove more reparable than those arising from the substance-related addictions, although empirical evidence is lacking for this hypothesis. In particular, neuroscientific studies are needed to directly contrast Internet addicts with substance dependent addicts in longitudinal designs.

From a therapeutic viewpoint, a large difference exists between Internet addiction and other forms of addiction. While typically a 'zero tolerance' strategy is taken toward future drug use with substance-abusing addicts, full abstinence from the Internet cannot be the aim in treating Internet addicts. Otherwise, most jobs and communication forms of the 21st century are out of reach for the afflicted patients. Interestingly, empirical data also suggests that only private but not business use of the Internet is associated with Internet addiction (Montag et al. 2010; Sariyska et al. 2014). On the other hand, it has been demonstrated that Internet addicts have already been successfully treated with classic psychopharmaceuticals, such as SSRIs (see review by Camardese et al. 2012, see also Chap. 10), showing that Internet addiction might not that different to treat, compared with other psychopathological disorders. Moreover, successfully treated Internet addicts have been associated with a down-regulation of activity in the dorsolateral PFC, potentially reflecting lower rumination about craving (Han et al. 2010; see also the Chap. 5). In sum, therapy seems to leave its mark on the activity of the brain as demonstrated with fMRI here.

8.7 Limitations of the Present Short Synopsis

Last but not least, some limitations of this very brief overview need to be mentioned. First of all, for reasons of clarity and to give an "easy to read" overview on the neuroscientific findings related to Internet addiction, we focused on two major brain regions in addiction research—namely the ACC and the striatum. Of course, a large number of other brain areas stemming from the limbic system and further prefrontal regions have been not included here. Second, the present overview followed not the aim to review all existing neuroscience studies on Internet addiction, because this has been done much better by the authors of each chapter and their special approach to Internet addition. Third, some new approaches are arising in neuroscience, which have not yet been implemented in Internet addiction research. As depicted in Fig. 8.2, epigenetics will be of interest to better understand how environmental influences changes genetic activity into the direction of becoming vulnerable or resilient for becoming addicted to the Internet.

References

Berke JD, Hyman SE (2000) Addiction, dopamine, and the molecular mechanisms of memory. Neuron 25:515–532

Blum K, Sheridan PJ, Wood RC et al (1996) The D2 dopamine receptor gene as a determinant of reward deficiency syndrome. J R Soc Med 89:396–400

Camardese G, De Risio L, Di Nicola M et al (2012) A role for pharmacotherapy in the treatment of "Internet Addiction". Clin Neuropharmacol 35:283–289

Carter BL, Tiffany ST (1999) Meta-analysis of cue-reactivity in addiction research. Addiction 94:327–340

De Ruyck K, Nackaerts K, Beels L et al (2010) Genetic variation in three candidate genes and nicotine dependence, withdrawal and smoking cessation in hospitalized patients. Pharmacogenomics 11:1053–1063

Dong G, Lu Q, Zhou H, Zhao X (2010) Impulse inhibition in people with Internet addiction disorder: electrophysiological evidence from a Go/NoGo study. Neurosci Lett 485:138–142

Dong G, Zhou H, Zhao X (2011a) Male Internet addicts show impaired executive control ability: evidence from a color-word Stroop task. Neurosci Lett 499:114–118

Dong G, Huang J, Du X (2011b) Enhanced reward sensitivity and decreased loss sensitivity in Internet addicts: an fMRI study during a guessing task. J Psychiatr Res 45:1525–1529

Forman SD, Dougherty GG, Casey BJ et al (2004) Opiate addicts lack error-dependent activation of rostral anterior cingulate. Biol Psychiatry 55:531–537

Franklin TR, Acton PD, Maldjian JA et al (2002) Decreased gray matter concentration in the insular, orbitofrontal, cingulate, and temporal cortices of cocaine patients. Biol Psychiatry 51:134–142

Franklin TR, Wang J, Sciortino N et al (2007) Limbic activation to cigarette smoking cues independent of nicotine withdrawal: a perfusion fMRI study. Neuropsychopharmacology 32:2301–2309

Han DH, Lee YS, Yang KC et al (2007) Dopamine genes and reward dependence in adolescents with excessive internet video game play. J Addict Med 1:133–138

Han DH, Kim YS, Lee YS et al (2010) Changes in cue-induced, prefrontal cortex activity with video-game play. Cyberpsychol Behav Soc Netw 13:655–661

Kerns JG, Cohen JD, MacDonald AW et al (2004) Anterior cingulate conflict monitoring and adjustments in control. Science 303:1023–1026

Kim SH, Baik SH, Park CS et al (2011) Reduced striatal dopamine D2 receptors in people with Internet addiction. Neuroreport 22:407–411

Ko CH, Liu GC, Hsiao S et al (2009) Brain activities associated with gaming urge of online gaming addiction. J Psychiatr Res 43:739–747

Koepp MJ, Gunn RN, Lawrence AD et al (1998) Evidence for striatal dopamine release during a video game. Nature 393:266–268

Li Q, Wang Y, Zhang Y et al (2012) Craving correlates with mesolimbic responses to heroin-related cues in short-term abstinence from heroin: an event-related fMRI study. Brain Res 1469:63–72

Lubman DI, Yücel M, Pantelis C (2004) Addiction, a condition of compulsive behaviour? Neuroimaging and neuropsychological evidence of inhibitory dysregulation. Addiction 99:1491–1502

Montag C, Bey K, Sha P et al (2014) Is it meaningful to distinguish between generalized and specific Internet addiction? Evidence from a cross-cultural study from Germany, Sweden, Taiwan and China. Asia-Pacific Psychiatry

Montag C, Jurkiewicz M, Reuter M (2010) Low self-directedness is a better predictor for problematic internet use than high neuroticism. Comput Hum Behav 26:1531–1535

Munafo MR, Matheson IJ, Flint J (2007) Association of the DRD2 gene Taq1A polymorphism and alcoholism: a meta-analysis of case–control studies and evidence of publication bias. Mol Psychiatry 12:454–461

Noble EP (2000) Addiction and its reward process through polymorphisms of the D_2 dopamine receptor gene: a review. Eur Psychiatry 15:79–89

Sariyska R, Reuter M, Bey K et al (2014) Self-esteem, personality and Internet addiction: a cross-cultural comparison study. Personality Individ Differ 61:28–33

Sun Y, Ying H, Seetohul RM et al (2012) Brain fMRI study of crave induced by cue pictures in online game addicts (male adolescents). Behav Brain Res 233:563–576

Teh LK, Izuddin AF, Fazleen HM et al (2012) Tridimensional personalities and polymorphism of dopamine D2 receptor among heroin addicts. Biol Res Nurs 14:188–196

Thanos PK, Volkow ND, Freimuth P et al (2001) Overexpression of dopamine D2 receptors reduces alcohol self-administration. J Neurochem 78:1094–1103

Thompson J, Thomas N, Singleton A et al (1997) D2 dopamine receptor gene (DRD2) Taql A polymorphism: reduced dopamine D2 receptor binding in the human striatum associated with the A1 allele. Pharmacogenet Genomics 7:479–484

Volkow ND, Fowler JS, Wang GJ (2003) The addicted human brain: insights from imaging studies. J Clin Invest 111:1444–1451

Vollstädt-Klein S, Wichert S, Rabinstein J et al (2010) Initial, habitual and compulsive alcohol use is characterized by a shift of cue processing from ventral to dorsal striatum. Addiction 105:1741–1749

Wise RA, Rompré PP (1989) Brain dopamine and reward. Annu Rev Psychol 40:191–225

Zhou Y, Lin FC, Du YS et al (2011) Gray matter abnormalities in Internet addiction: a voxel-based morphometry study. Eur J Radiol 79:92–95

Part III
Therapeutical Interventions & Governmental Policies

Chapter 9
The Impact of Psychoinformatics on Internet Addiction

Christian Montag, Martin Reuter and Alexander Markowetz

Abstract *Psychoinformatics* refers to the new collaboration between the disciplines computer science and psychology to study psychological phenotypes by means of data mining. The present chapter gives an overview of how *Psychoinformatics* can aid research and therapy in the context of Internet addiction.

9.1 Introduction

Psychoinformatics, the fusion of psychology with computer science, will be of great importance in the treatment, diagnostics, and research of Internet addiction. In a recent paper, Markowetz et al. (2014) outline the potential of *Psychoinformatics*, as a new interdisciplinary research endeavor, for the study of mental health. This approach applies methods from computer science to psychological research, in order to obtain deeper insights into the mental states of individuals or other psychological variables such as personality (for an example for empirical research in this field see Montag et al. 2014). To date, the most frequently used methods in psychology for the assessment, understanding and prediction of human behavior, have been (i) the classic laboratory experiment, (ii) the administration of self-report questionnaires and (iii) particularly in psychotherapeutic and work psychology settings, the interview. Although these techniques have several advantages, they also have some drawbacks.

C. Montag (✉)
Department of Psychology, University of Ulm, Ulm, Germany
e-mail: christian.montag@uni-ulm.de

M. Reuter
Center for Economics and Neuroscience, University of Bonn, Bonn, Germany

A. Markowetz
Department for Informatics, University of Bonn, Bonn, Germany

M. Reuter
Department of Psychology, University of Bonn, Bonn, Germany

C. Montag and M. Reuter (eds.), *Internet Addiction*, Studies in Neuroscience, Psychology and Behavioral Economics, DOI 10.1007/978-3-319-07242-5_9

First, we briefly summarize the advantages and limitations of traditional methods in psychological research. Subsequently, we outline techniques from computer science that can address these issues. Finally, we outline the impact of *Psychoinformatics* on the research and therapy of Internet addiction.

9.2 Traditional Methods in Psychology

The interview poses one of the leading tools to obtain insights into the attitudes and character of a person (e.g. Potter and Hepburn 2005; Schmidt and Hunter 1998). It is not only used widely in psychotherapy, but in many other areas such as work and organizational psychology. The information gathered is used to hire a person for a job or to help a person in overcoming biased cognitions such as being always responsible for a negative outcome while treating depression (Wright and Beck 1983; Beck 2002).

Significant research effort has been directed into the development of structured interviews, to enable a fair hiring process (Campion et al. 1988) or to form guidelines for a psychiatric diagnosis, such as depression (Riskind et al. 1987). Structured interviews follow a clear 'structure', with the same items/questions being administered in the same order. In contrast unstructured interviews follow no clear line of questioning. The push towards standardized interviews has led to improved outcomes with respect to their validity, e.g. in work psychology (Wiesner and Cronshaw 1988) and clinical psychology (Miller et al. 2001). The enduring prevalence of interviews in many areas of psychology can be attributed to the fact that direct communication with a person is, in many ways, an indispensable tool (especially in therapy), in taking into account the unique perspective of a person. This subjective perspective can be supplemented with information derived from other (more objective) sources. Imagine a salesperson who pretends to be outstanding at selling cars. In reality, it turns out that, despite the self-report, he/she has never sold a car. The discrepancy between the information derived from the interview and the exact figures from a company's paperwork on sold cars, gives insights into biased cognitions. In the context of the main topic of the present chapter it is noteworthy that Tao et al. (2010) investigated the prevalence of different diagnostic criteria for Internet addiction; which represents a significant step towards a structured interview (as questions can be built based on these findings). Notable is their proposal of a '2 + 1 rule'. As a first step, the symptoms 'preoccupation with the Internet' and 'withdrawal' must be observed. Adding to this, at least one out of several optional symptoms, e.g. development of tolerance, must also be present, in order to justify a diagnosis of Internet addiction.

Besides the mentioned advantages of interviews, this method also has several shortcomings. First, the method is time consuming and expensive, requiring the psychologist as well as the interviewee to invest a significant amount of time in data collection. Second, it is unclear whether the interviewee provides an accurate description of his/her mental world. Social desirability may play a particularly important role, especially where individuals are asked about stigmatizing psychopathological conditions (Edwards 1957; Sugarman and Hotaling 1997). Social desirability describes the tendency to present one's own character and attitudes according to what is thought to be acceptable in terms of social norms. Usually the person describes him- or herself as 'better' than he/she actually is. However, the opposite, so-called "self-handicapping", is also observed. Here, individuals 'handicap' themselves to protect their self-esteem, e.g. if a goal could not be achieved. It is important to note that cross-cultural differences in tendencies toward social desirability also exist (Johnson and van de Vijver 2003).

The second dominant method in the social sciences is the use of self-report-questionnaires. Here, a person fills in items on a given topic of interest—e.g. in this instance, Internet addiction. Although questionnaires are relatively easy to handle, they also need to be analyzed (i.e. through use of statistical tests) and they too are subject to the problem of social desirability. This is especially true if the information provided on the questionnaire is not anonymized, but needs to be used further, e.g. in a therapeutic or employment context. Besides this, classic forms of self-report-inventories, such as paper-pencil questionnaires, constitute extra work for the researcher, as the information must be transferred to a digital form (to enable statistical analysis) and the hard-copy questionnaires must then be stored for a longer time. This is especially true where paper-pencil questionnaires have been part of a research endeavor, employment scenario, etc. Of course, this can be circumvented through use of electronic versions of questionnaires, whereby some of the aforementioned problems can be diminished. Yet, the major drawback of self-report methods and interviews remain; the reliance upon the ability of the participant to recognize, recollect, and present the subject matter accurately.

Last but not least, social scientists rely on experiments, in which strict experimental laboratory settings provide the possibility of controlling potential confounding factors, e.g. if a researcher aims to assess the influence of Internet addiction on cognitive function. Although experiments provide a high scientific standard, they are vulnerable to several problems. Due to the artificial environment of the laboratory, it is not clear whether results derived from laboratory experiments can be generalized to the 'real world'. In other words laboratory experiments often lack ecological validity, especially when only a small aspect of a broad phenotype is assessed in the lab. For example, shorter fixation times to threatening stimuli (a marker for anxiety) do not imply that a participant has difficulties in giving a talk in front of a broad audience (social phobia). Besides this, and in-keeping with the other methods, participants' motivational factors play an important role. If a participant in a study is not motivated to follow the instructions of the experiment, the quality of the data will be poor. Moreover, conducting experiments is very expensive, because a room for the experiment needs to be rented and highly trained specialists are needed to carry out the experiment properly.

Methods from computer science can help to overcome some of the above-mentioned problems encountered by interviews, self-report-measures and experiments. They can be of particular value to research efforts in Internet addiction.

9.3 Psychoinformatics and Internet Addiction

Methods from computer science have been used in psychological research for several decades. The use of computers and the Internet, such as online surveys and computer based psychological experiments, illustrate this point nicely. Such, rather obvious, methods will not be discussed in the present chapter. Instead, we focus on new developments in the assessment of human behavior, via mobile devices but also by monitoring the human–machine interaction in everyday life e.g. the use of the computer in a workplace situation.

Most humans in industrialized societies possess and use a smartphone on a daily basis. These powerful little machines provide access to the Internet, entertainment, including computer games, and of course the classic functions of mobile phones, such as telephone and short text message features. Furthermore, the new generation of smartphones includes intelligent sensors, which are able to track the location of a person (via the global positioning system, GPS) and their bodily movements. Of note, the smartphone is usually carried on the body of a person, thereby accompanying humans in nearly all daily activities. Many smartphone users even carry their smartphone when going for a run to track the speed and distance of the workout. This being said, smartphones and the data recorded from the human-machine interaction, provide genuine insights into the life and even mental states of a person. Most importantly, data derived from this source can be collected on a longitudinal basis, whereas the traditional methods discussed above usually do not go beyond one or two measures. Anyway, traditional methods do not provide a fine grained monitoring of a person's behavior over a long time.

In the context of assessing and treating Internet addiction, an exact recording of the length of online sessions over a longer time window provides a significantly more accurate picture of a person's Internet usage than numbers derived via self-report. Direct monitoring of the smartphone also provides insight into the most important addictive behaviors in the context of computer or smartphone use. The need to distinguish between different forms of online addiction has been demonstrated recently by Montag et al. (2014).[1] Figure 9.1 depicts our recently self-developed app for monitoring smartphone behavior.

By tracking Internet activity directly on a smartphone or computer, classic symptoms from addiction research can be examined. If one thinks of development of tolerance in the context of Internet consumption, this symptom should be reflected in increasing hours spent online over a given time window. Of course, here a patient's wellbeing will also need to be considered, and will play an important role in diagnosing Internet addiction. The therapist's impression of a patient completes the assessment of this potential new behavioral addiction. Besides

[1] Of note, smartphone addiction (and the broader category mobilephone addiction) and Internet addiction are not the exact same constructs, as questionnaires measuring both forms of addiction correlate only moderately with each other (see Jenaro et al. 2007). But: As a rising number of smartphone users surf the Internet via these devices, it is very likely that correlations will be higher in the future.

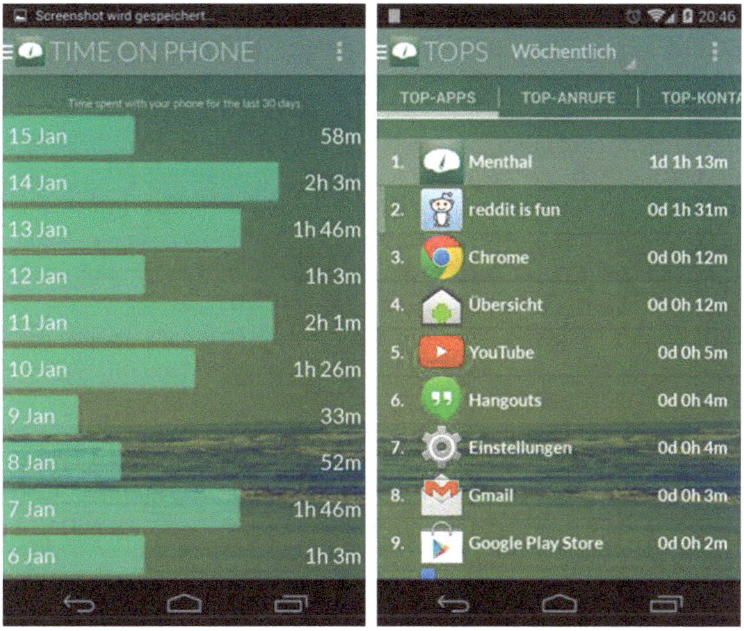

Fig. 9.1 *Left side* of the figure shows the recording of actual time spent on the smart phone each day and the *right side* shows the most often used application of a user. The figure depicts *screenshots* from the app "Menthal"

asking a person about his/her wellbeing, it will be also possible to assess wellbeing via *textmining* in the future. Here, the content of e-mails, messages of online social network channels, etc. can be analyzed in terms of the number of positive and negative words used by the patient. This will give indirect insights into the individual's mental state. This topic will prove less intrusive than anticipated, as the text can be analyzed on the phone and only the derived quantitative assessment is transmitted to a server (not the content of an individual's messages).

In the following section we outline a short example demonstrating how a distinct pattern of different variables from smartphone use, could help to monitor and diagnose affective disorders such as depression. Depression is of interest in the context of Internet addiction as a subgroup of Internet addicts concurrently experience a form of this affective disorder (Kim et al. 2006). Imagine that a person usually phones ten different contacts a day; this person is very active in terms of recorded GPS locations, and shows signs of positive emotionality reflected in the large number of positive words used in communication channels (including lots of smiley emoticons). Suddenly a different data pattern occurs: The same person calls no one for several weeks, seems to stay at home (no GPS activity) and the use of negative words (and a lower number of messages) prevails. The shift in the data pattern recorded on the smartphone could represent a sign of social withdrawal accompanied by negative emotionality, possibly indicating a depressive state.

Psychoinformatics can also aid in the therapy of Internet addicts. Traditional settings such as Cognitive Behavioral Therapy (CBT) usually encounter longer time windows between therapy sessions. Here, the therapist often sees his/her patient not more than once a week. In addition, in some psychotherapeutic approaches (see also the Chaps. 11 and 12), patients are required to write a daily diary monitoring their Internet usage and/or their mental states. The inclusion of the above-mentioned techniques from *Psychoinformatics* aids and simplifies the therapeutic process, because events in the everyday life of a patient can be better included in therapy, without the problem of distorted memory. For example, if the therapist verbally asks the patient "How often have you used the Internet over the last week in hours?", this question can only be answered with a vague number. It might even be trickier if the patient is required to recall the use of last Monday. In contrast to this, direct measurement of time spent using the device will provide accurate numbers and facilitate more direct therapeutic interventions. If an Internet addict uses a certain function of the Internet, such as Facebook, for more than one hour, a virtual 'red flag' could be raised alerting the user to quit the session. It is even possible that after excessively long online sessions a notification will be send directly to the therapist.

Last but not least, *Psychoinformatics* can help to assess the development of cognitive function in Internet addicts. Among others, Park et al. (2011) provided evidence that Internet addiction is associated with lower cognitive functions in the domain of attentional processes. As this data stems from a cross-sectional study, longitudinal evidence is needed to obtain insights into the cause-effect principles. Again, the smartphone may be of help. Instead of 'swiping' the log-in-screen to gain access to the functions of a smartphone, cognitive games can be incorporated at log-in (e.g. instead of swiping, the patient participates in a short cognitive test). Thousands of data points on cognitive functions could thus be collected by the therapist or the researcher, giving valuable insights into the development of cognition in light of Internet and smartphone use. Naturally, it is not practical to include a five minute experiment at log-in, but this may be viable for shorter trials (such as a ten second game when logging in). Longer experiments would possibly serve to demotivate participants' smartphone usage (albeit arguably show some therapeutic effects), but also could demotivate participants' participation in the experiment. Although no data yet exist on the validity and reliability of cognitive data collected via a smartphone log-in-screen, clearly the collection of thousands of data points should drastically diminish the standard error of the mean in the registered data and therefore provides a more precise picture of the user's smartphone usage cognitive functions.

9.4 Psychoinformatics, Multiple Testing and Data Privacy

The use of *Psychoinformatics* also presents several problems. First of all, we would like to consider the important issue of data protection/privacy. The recording of data from smartphones or from other digital devices bears the great danger of data misuse. Therefore, it is of importance to explain in detail which data will be recorded, and what kind of data will *not* be recorded. In the above-mentioned 'Menthal Balance' app (Fig. 9.1), users are informed prior to installation, that no calls or message content will be recorded. Furthermore, in this research scenario, each smartphone is used as a small computer to conduct the statistical analyses. Only numbers, i.e. no direct content, from the smartphone is sent to the server of the researchers. Nevertheless, for many areas, such as therapy, direct content from mails could be a valuable data source. Again, we wish to note the difference between categorizing words into positive or negative categories and sending numerical data on the use of these positive/negative words to a server, compared with sending the complete content. Of note, the long recognized system of confidentiality between medical/psychological practitioners and patients provides a valuable guideline on how to use the data derived from *Psychoinformatics*.

A second problem of *Psychoinformatics* applies primarily to research related areas. Correlating thousands of variables from Big Data[2] will inevitably lead to some false positive results. Therefore, strong correction procedures, such as those applied in genetics and brain imaging research areas need to be adapted in this kind of research approach. Results derived from exploratory data analyses will also need to be replicated.

9.5 Conclusions

Psychoinformatics, the introduction of methods from computer science to the area of psychology, can aid the diagnostics process, therapy, and also research in the field of Internet addiction. To date, there is a dearth of empirical data in support of the above-mentioned hypotheses. The future will show how these methods can be included in the various areas dealing with Internet addiction.

References

Beck AT (2002) Cognitive models of depression. Clin Adv Cogn Psychother Theor Appl 14:29–61

Campion MA, Pursell ED, Brown BK (1988) Structured interviewing: raising the psychometric properties of the employment interview. Pers Psychol 41:25–42

Edwards AL (1957) The social desirability variable in personality assessment and research. Dryden Press, USA

Jenaro C, Flores N, Gómez-Vela M et al (2007) Problematic internet and cell-phone use: Psychological, behavioral, and health correlates. Addict Res Theor 15:309–320

Johnson TP, van de Vijver FJ (2003) Social desirability in cross-cultural research. Cross Cult Surv Methods 195–204

[2] In our context, Big Data refers to the enormous amount of information being collected on devices such as a smartphone.

Kim K, Ryu E, Chon MY et al (2006) Internet addiction in Korean adolescents and its relation to depression and suicidal ideation: a questionnaire survey. Int J Nurs Stud 43:185–192

Markowetz A, Błaszkiewicz K, Montag C et al (2014) Psycho-Informatics: Big Data shaping modern psychometrics. Med Hypotheses 82:405–411

Miller PR, Dasher R, Collins R et al (2001) Inpatient diagnostic assessments: 1. Accuracy of structured vs. unstructured interviews. Psychiatry Res 105:255–264

Montag C, Bey K, Sha P et al (2014) Is it meaningful to distinguish between generalized and specific Internet addiction? Evidence from a cross-cultural study from Germany, Sweden. Asia-Pacific Psychiatry, Taiwan and China

Montag C, Błaszkiewicz K, Lachmann B et al. (accepted) Correlating personality and actual phone usage: evidence from psychoinformatics. J Individ Differ

Park MH, Park EJ, Choi J et al (2011) Preliminary study of internet addiction and cognitive function in adolescents based on IQ tests. Psychiatry Res 190:275–281

Potter J, Hepburn A (2005) Qualitative interviews in psychology: problems and possibilities. Qual Res Psychol 2:281–307

Riskind JH, Beck AT, Berchick RJ et al (1987) Reliability of DSM-III diagnoses for major depression and generalized anxiety disorder using the structured clinical interview for DSM-III. Arch Gen Psychiatry 44:817

Schmidt FL, Hunter JE (1998) The validity and utility of selection methods in personnel psychology: practical and theoretical implications of 85 years of research findings. Psychol Bull 124:262

Sugarman DB, Hotaling GT (1997) Intimate violence and social desirability a meta-analytic review. J Interpersonal Violence 12:275–290

Tao R, Huang X, Wang J et al (2010) Proposed diagnostic criteria for internet addiction. Addiction 105:556–564

Wiesner WH, Cronshaw SF (1988) A meta-analytic investigation of the impact of interview format and degree of structure on the validity of the employment interview*. J Occup Psychol 61:275–290

Wright JH, Beck AT (1983) Cognitive therapy of depression: theory and practice. Psychiatric Serv 34:1119–1127

Chapter 10
Pharmacological Treatment of Internet Addiction

Giovanni Camardese, Beniamino Leone, Coco Walstra, Luigi Janiri and Riccardo Guglielmo

Abstract The increasing number of Internet users has resulted in an increased population percentage affected by the negative effects of problematic Internet usage. To date, the management of psychopathological Internet use is not supported by extensive empirical research. No standard clinical treatment protocols for pharmacological treatment exist, and as a result, empirical or anecdotal assessments based on case studies are mainly consulted. A relevant problem in performing clinical trials is the evolving nosology, which encompasses ambiguous definitions of Internet addiction and a diversity of diagnostic, prognostic, and therapeutic criteria. The aim of this chapter is to review the current literature, to assess the extent to which specific pharmacological interventions (e.g., using antidepressants, mood stabilizers, opioid receptor antagonists, or antipsychotics) can alleviate the symptomatic burden in patients with "Internet addiction." We also explore pharmacological interventions that target patterns of comorbidity and underlying psychopathological dimensions (e.g., addiction, impulsivity, obsessive-compulsive spectrum, bipolar spectrum, dissociation, etc.) shared with other behavioral or substance addictions.

10.1 Introduction

The Internet represents one of the most important products of culture and industry in society and has become an integral part of daily life for many people. Worldwide, more than one billion computers are connected to the Internet (Reuter et al. 2005) and the increasing number of Internet users has, in fact, resulted in an increased percentage of users being afflicted by problematic Internet use. This phenomenon is growing, both in terms of prevalence and within the public consciousness, as a pathological

G. Camardese (✉) · B. Leone · C. Walstra · L. Janiri · R. Guglielmo
Institute of Psychiatry and Psychology, Catholic University of Sacred Heart,
L.go A. Gemelli 8, 00168 Rome, Italy
e-mail: g.camardese@rm.unicatt.it

C. Montag and M. Reuter (eds.), *Internet Addiction*, Studies in Neuroscience, Psychology and Behavioral Economics, DOI 10.1007/978-3-319-07242-5_10

condition. This societal development together with an increase in clinical observations, raise issues concerning the management of the condition. To date, there is no consensus in the literature with respect to the definition of problematic Internet use, though most authors refer to it as "Internet Addiction" (IA) (Tao et al. 2010).

The study of IA is currently hampered by ambiguous definitions of the phenomenon and a diversity of diagnostic, prognostic and therapeutic criteria. Internet use can lead to a state that appears to meet the Diagnostic and Statistical Manual of Mental Disorders (DSM) definition for a mental disorder, described as "a clinically significant behavioral or psychological syndrome associated with present distress or with a significantly increased risk of suffering death, pain, disability or an important loss of freedom" (APA 2013). No clinical conditions related to problematic Internet use have been part of any diagnostic system, until now. Considerable effort has been made to include "Internet addiction", "pathological Internet use", "problematic Internet use" (Shapira et al. 2000) or any of its derivatives in the 2013 update of the Diagnostic and Statistical Manual of Mental Disorders, the Fifth Edition (DSM-5), but no official criteria exist in either the DSM-5 or in the International Classification of Diseases, Tenth Revision (WHO 2010). To date, only "Internet gaming disorder" has been included in Sect. 10.3 of DSM-5 (American Psychiatric Association 2013), and the core feature of this disorder is the persistent and recurrent participation in computer gaming, typically massively multiplayer online role play games, for many hours. This condition is included to reflect the scientific literature (most of which comes from studies in Asian countries) on persistent and recurrent use of Internet games, and preoccupation with these, which can result in clinically significant distress and functional impairment of general life, such as social interaction, academic performance, occupational interest, and behavioral problems (Petry and O'Brien 2013). As a syndrome listed in Sect. 10.3 of DSM-5, "Internet gaming disorder" requires further research before it can be formally considered as a disorder in its own right.

Although the diagnostic criteria and assessment questionnaires used for diagnosis may vary between countries, surveys in the United States and Europe indicated that IA may affect between 1.5 and 8.2 % of the general population (Weinstein and Lejoyeux 2010). To date, clinicians have only empirical or anecdotal reports at their disposal, concerning pharmacological treatment options for the management of the large number of patients suffering from IA. For this reason, in our chapter we review the role of pharmacotherapy in the treatment of IA to guide clinical decisions according to the most recent data. We also provide a psychopathological framework for this particular form of behavioral addiction, with the intention of proposing a guided approach to aid clinicians in choosing between available drugs.

10.1.1 Is Pharmacological Treatment a Valid Therapeutic Option for IA?

In order to provide an answer to this question, an overview of the empirical data on functional changes occurring in the brains of patients suffering from non-substance related addictions is of relevance. The identification of specific neurophysiological dysfunction in behavioral addictions would establish a biological rationale for intervention. Data to this effect are scarce in the literature.

The potential phenomenological overlap between IA, substance addiction and, in particular, gambling suggests that a common neurobiological substrate involving an impairment of the "reward system" underlies these disorders. The mesolimbic dopaminergic pathway represents the final common pathway for reinforcement/reward induced by physiological stimuli or psychotropic drugs. It follows that dopamine is considered the neurotransmitter responsible for mediating "pleasure" (Di Chiara and Bassareo 2007). The intake of certain substances or the execution of certain behaviors, induces a very intense and fast feeling of pleasure (consummatory pleasure) caused by the rapid increase of dopamine in the mesolimbic system (Di Chiara and North 1992). Moreover, dopamine is also elevated when anticipating the substance use or behavioral execution (anticipatory pleasure) and this also drives motivational processes that promote goal-directed behaviors aimed at achieving desired rewards. Addiction disrupts the normal activity of these dopaminergic circuits, thus redefining the hierarchy of motivational priorities.

In this regard, the literature provides small, but convincing evidence for a link between biological brain abnormalities in patients addicted to substances and similar brain abnormalities in patients with IA. Blum et al. (2012) persuasively linked a reward-deficient aberrant behavior (RDAB) to abnormal dopaminergic function in the nucleus accumbens, and argued that RDABs can be observed both in conventional substance-use disorders, and also in excessive internet gaming and related activities that stimulate excessive dopamine release, such as gambling.

Several fMRI studies in pathological gamblers have reported blunted neural responses to appetitive cues, primarily in ventral striatum and orbital/lateral prefrontal cortex (Reuter et al. 2005; de Ruiter et al. 2009; Balodis et al. 2012) and these observations have been interpreted in terms of the reward deficiency hypothesis of addiction.

Furthermore, a neuroimaging study conducted by Ko et al. (2009) suggested that both the strong desire to play online video games[1] and craving e.g. for nicotine/alcohol in substance dependent addictions could be explained by a single neurobiological mechanism. The right orbitofrontal cortex, the medial frontal cortex, the anterior cingulate cortex bilaterally, the right dorsolateral prefrontal cortex, the caudate nucleus and the right nucleus accumbens were activated when patients were stimulated with images of "online gaming" in contrast to a neutral picture condition. These imaging results suggest that "online gaming addiction" indeed shares biological substrates with substance addiction. For instance, the urge to smoke cigarettes while watching a videotape showing smoking scenes in current smokers was associated with increased metabolic activity in the ventral striatum, anterior cingulate, orbitofrontal cortex, middle temporal lobe, hippocampus, insula, midbrain and thalamus (Weinstein et al. 2010).

[1] Characterizing a particular form of IA, namely "online game addiction".

From this it could be intuitively derived that IA might be treatable in the same way as other addictions. In 2010, Liu et al. (2010) carried out a functional magnetic resonance imaging (fMRI) experiment, by using the regional homogeneity (ReHo) method to analyze cerebral function characteristic of IAD college students under resting state. In adolescents with IAD, compared to healthy controls, they found that the increased ReHo brain regions (representing the increase in cerebral metabolic rate) were distributed over the cerebellum, brainstem, limbic lobe and frontal cortex, indicating a possible involvement of the "reward" system.

Summing up the evidence, to date the knowledge on the neurobiological underpinnings of IA is extremely limited and is insufficient as a basis for pharmacological intervention. Nevertheless, if we assume that a malfunction of the reward system underlies IA, one might conclude that pharmacological interventions of use in treating other forms of addiction may be eligible as a starting point for psychopharmaceutical research in the area of IA.

10.2 Clinical Evidence on Pharmacotherapy for IA

To date, case studies of IA treatment are rather scarce and several key limitations have been highlighted, including inconsistencies in definition and diagnosis, a lack of randomization and blinding techniques, a lack of adequate controls or other comparison groups, and insufficient information concerning recruitment dates, sample characteristics, and treatment effect sizes (King et al. 2011).

First, we reviewed the current literature on pharmacotherapy specifically for "Internet addicted" patients. However, considering the lack of adequately large, rigorous studies, we also focused, in a second step, on the underlying psychopathological dimensions of IA (i.e., impulsivity, compulsivity, craving, obsessive-compulsive spectrum) as well as on the high prevalence of comorbid conditions to address the issue of dual diagnosis.

We discuss the clinical evidence available for different classes of psychotropic drugs, according to our recent article (see Camardese et al. 2012).

10.2.1 Antidepressants

Antidepressants are psychotropic drugs mainly used in depression, dysthymia and anxiety. They include: monoamine oxidase inhibitors, tricyclics, selective serotonin reuptake inhibitors (SSRIs), serotonin-norepinephrine reuptake inhibitors and melatonergic agents. The onset of the antidepressant effect is delayed (4–8 weeks) and treatment typically lasts for months or years. They determine mood enhancement, alertness and attention, increased appetite, regularization of sleep and reduction of the hypochondriac attitude. Moreover, serotonergic antidepressants have been proven to be helpful to alcoholics in maintaining abstinence and decreasing craving for alcohol in detoxified alcohol dependent subjects (Janiri et al. 1998). Bupropion (a dopamine/norepinephrine reuptake inhibitor) is also effective for smoking cessation (Hughes et al. 2007).

The use of antidepressants for IA, in particular SSRIs may be endorsed mostly by evidence of the aminergic systems' role in the suppression of inhibitory control (i.e. "resisting" the urge) and the control of compulsive repetition, as well as data indicating a high lifetime prevalence of major depression in "internet addicts" (Shapira et al. 2003; Yen et al. 2007; Lee et al. 2008). Clinical studies have also suggested a close relationship between serotonergic dysregulation, impulsivity, and symptoms of the obsessive-compulsive spectrum, for which serotonergic drugs are known to be effective (Goddard et al. 2008). However, while definitely effective in treating obsessive-compulsive disorder, SSRIs have shown mixed results in some impulse control disorders, namely pathological gambling, kleptomania and compulsive shopping (Kim et al. 2002; Koran et al. 2002, 2007; Grant et al. 2009).

The first experience reported in the literature on IA concerns treatment with escitalopram (a SSRI) in an "internet gaming addicted" patient (30 mg/day for 3 months). An improvement in mood and a significant reduction of the strong urge to perform online gaming were observed, leading to a complete functional recovery (Sattar and Ramaswamy 2004). Another study investigated the effectiveness of escitalopram (20 mg/day for 10 weeks) on 19 "Internet addicts" (Dell'Osso et al. 2008). During the 10 week open-label phase,[2] 11 patients (64.7 % of the sample) showed significant decreases in weekly hours spent online and improvements in global functioning. At the end of the 10 weeks, subjects were blindly randomized to either continued escitalopram treatment or to placebo. The abovementioned improvement persisted in the second phase of the study, but no significant differences were observed between those who continued taking the drug and those who were switched to placebo. The authors speculate that 9 weeks may not have been sufficient for the effect to be lost in the placebo group, or for additional gains to be made in the escitalopram group, but also do not rule out the possibility that the improvements seen in the open-label phase may have been a placebo response.

A study on 11 "Internet game addicted" patients and 8 healthy controls, assessing brain activity in response to a game stimulus using functional magnetic resonance imaging (fMRI), explored the possible effectiveness of bupropion, a norepinephrine/dopamine reuptake inhibitor. Patients showed higher activation in the left occipital lobe, left dorsolateral prefrontal cortex and left parahippocampal gyrus before treatment. After 6 weeks of treatment, craving for Internet video game play, total game play time, and cue-induced brain activity in the dorsolateral prefrontal cortex were decreased (Han et al. 2010). This allowed the authors to presume that the drug was effective but, given the small sample size, further studies are needed to support this assumption.

A study conducted on Internet addicts with comorbid major depressive disorder and excessive online game play, further investigated the role of bupropion in

---

[2] A clinical trial where both physician and patient know about the administered drug.

reducing the severity of online game play as well as depressive symptoms (Han and Renshaw 2011). The study consisted of a 12-week, randomized, double-blind clinical trial, including an 8-week active treatment phase and a 4-week post-treatment follow-up period. Significant bupropion-associated reductions in online gaming and depressive mood were observed, though the latter improvement did not persist during the post-treatment follow-up period.

The anti-craving properties of antidepressants still need to be evaluated with respect to long-term outcomes and in controlled studies. Regarding observations on anti-craving properties made in other addiction research, evidence for short-term effects seems to exist in the main. Long-term exposure to antidepressants may also facilitate mood swings toward the manic pole, to which subjects with pathological addictions seem more prone (Goldberg and Whiteside 2002). This could imply a greater risk of relapse of impulsive behavior, which is characteristic of manic mood.

10.2.2 Opioid Receptor Antagonists

Opioid receptor antagonists such as naltrexone and nalmefene block the reinforcing effects of opioids and reduce substance consumption and craving. Generally, they have no abuse potential, mild and transient side effects, and appear to be a suitable treatment for addiction in combination with psychosocial support. Opioid receptor antagonists are mainly prescribed in alcoholism and heroin dependence.

Various studies have found a high comorbidity rate between IA and other forms of addiction (such as a substance use disorder) as well as impulse control disorders (Bernardi and Pallanti 2009). In this regard, Griffiths (2000) suggested that the Internet merely represents a different context in which gamblers, shopping addicts and sex addicts develop their pathological behavior. In particular, Davis (2001) distinguished a "specific pathological Internet use" and a "generalized pathological Internet use". The first includes online sexual material/services, online auction services, online stock trading and online gambling. These dependencies are content-specific and they would exist in the absence of the Internet and in a manner independent of multiple Internet functions. The second involves a general, multidimensional overuse of the Internet and it may be related to the social aspect of the Internet. Individuals with general pathological Internet use are considerably more problematic, in that their pathology would likely not even exist in the absence of the Internet (Davis 2001).

In our opinion, the mounting evidence on pathological gambling considered by the authors could equally be applied to some IA patients, given that gambling shares, as previously mentioned, the conceptual and phenomenological bases of a behavioral addiction. The psychopathological (e.g. impulsivity, compulsive repetition, etc.) overlap between IA, substance addiction and pathological gambling suggests a common neurobiological substrate, involving a dopamine dysfunction of the neural "reward systems". Opioid receptor antagonists inhibit dopamine release in the nucleus accumbens and ventral pallidum and have been considered for use in some behavioral addictions. Literature is currently limited to a single case report in the treatment of IA with opioid receptor antagonists (see Bostwick and Bucci 2008). This case study reported successful treatment with naltrexone, which has also proven effective in the treatment of other impulse control disorders, such as pathological gambling and kleptomania (Grant et al. 2008, 2009). The patient in this study was a 31-year-old male with compulsive cybersexual behavior (chatting online, masturbating for hours, and occasionally, sex with Internet contacts). A stable dose of sertraline was ineffective in treating his "Internet addiction". Naltrexone (150 mg/day) was gradually administered and helped to induce a 3-year remission. The authors hypothesize that by blocking the capacity of endogenous opioids to trigger dopamine release in response to reward, naltrexone may block the reinforcing nature of compulsive Internet sexual activity.

10.2.3 Mood Stabilizers

Mood stabilizers are drugs that have the property of acting on mood. They affect long-term mood stability and typically lead to an improvement of the initial condition (e.g. depression, anxiety, agitation etc.). Mood stabilizers, such as lithium or anticonvulsants (e.g. valproic acid, carbamazepine and lamotrigine), are primarily used to treat bipolar disorder (BD). On the other hand, non-mood stabilizing anticonvulsants (e.g. gabapentin and pregabalin) are also increasingly used to treat alcoholism and substance abuse disorders. Generally, they show a safe side-effect profile and are well tolerated by patients (Guglielmo et al. 2012).

A potential use for mood stabilizers in the treatment of IA may be substantiated by similarities between mood disorders belonging to the bipolar spectrum and IA. Both conditions are characterized by impulsive behaviors (mainly during manic episodes in bipolar patients) and often coexist. In particular, a high lifetime prevalence (up to 70 %) of bipolar disorder has been found in Internet addicted patients. (Shapira et al. 2003; Di Nicola et al. 2010b).

A specific anti-compulsive property of some mood stabilizers has been hypothesized. Lithium and anticonvulsants have, in fact, been successfully used in the treatment of various impulse control disorders (Roncero et al. 2009). Likewise, in patients with substance use disorders, valproate appears to be a potentially fruitful medication due to its anticraving property (Maremmani et al. 2010). Furthermore, there is data pointing to the considerable utility of mood stabilizers (particularly lithium and valproate) in the treatment of pathological gambling (Pallanti et al. 2002). At present, the effectiveness of mood stabilizers in the treatment of IA has not been investigated, though consideration of this drug class in future studies is certainly promising.

10.2.4 Antipsychotics

Antipsychotics include drugs used for the treatment of psychotic diseases, such as schizophrenia or bipolar disorder, mainly acting on neurotransmitter systems of dopaminergic and serotonergic pathways. Antipsychotics can be divided in first-generation antipsychotics (typical antipsychotics or neuroleptics) and second-generation antipsychotics (atypical). The main difference is that typical antipsychotics have a highly selective affinity for D2 receptors while atypical antipsychotics can modulate both dopaminergic and serotonergic systems in different ways. Due to their relevant action on dopamine receptors of the nigrostriatal pathway, neuroleptics increase the risk of extrapyramidal effects, unlike atypical antipsychotics, which are more tolerable in terms of neurological side-effects. On the other hand, due to their particular pharmacodynamic properties, atypical antipsychotics are also frequently used in other clinical conditions like mood disorders, anxiety, and autism spectrum disorders.

With respect to these drugs, possible models for their use in patients with addiction are linked with antipsychotics' effectiveness in the treatment of resistant obsessive-compulsive disorder (Choi 2009). Given their serotonergic properties, atypical antipsychotics have been most investigated. Prescription of atypical antipsychotics seems to be a highly helpful strategy for treatment-resistant obsessive-compulsive disorder, with benefits most evident for risperidone (Bloch et al. 2006). Authors of several placebo-controlled clinical trials have found evidence to support psychopharmaceutical treatment with olanzapine (Bystritsky et al. 2004), risperidone (Hollander et al. 2003) and quetiapine (Denys et al. 2004). Head-to-head comparisons involving these agents have also been conducted: Maina et al. (2008) compared olanzapine and risperidone augmentation in subjects resistant to SSRIs and found that both were equally effective at reducing obsessive-compulsive symptoms. In particular, medication augmentation refers to the addition of a second drug to an initial, ineffective pre-existing therapy. A single pilot trial of atypical antipsychotic monotherapy using aripiprazole has also been published (Connor et al. 2005), with significant improvement observed, particularly in compulsive symptoms. Moreover, preliminary data also support a possible efficacy of aripiprazole in reducing alcohol craving (Martinotti et al. 2007, 2009).

The use of antipsychotics in treating impulse control disorders has also been investigated, given that the central features 'impulsivity' and 'compulsive repetition' are possible targets for antipsychotic medication. In particular, preliminary studies have shown that olanzapine, targeting both dopaminergic and serotonergic functioning, effectively reduces impulsivity. Olanzapine has shown preliminary effectiveness in several disorders in which a lack of impulse control is a key feature, such as trichotillomania, skin picking, and borderline personality disorder (Garnis-Jones et al. 2000; Stewart and Nejtek 2003; Christensen 2004; Shoja-Shafti 2006). Each of the clinical conditions that responded to olanzapine share phenomenological features with pathological gambling, in that patients are unable to resist impulses, and act without thinking about the consequences. Olanzapine has, thus, been tested in the treatment of pathological gambling, though it did not show significant effectiveness (McElroy et al. 2008). Quetiapine has also been tested as a treatment of pathological gambling and, in addition, it has been used as an add-on treatment for the management of bipolar I disorder with comorbid compulsive shopping and physical exercise addiction (Di Nicola et al. 2010a).

Given the overlap between impulsivity/compulsivity symptoms of IA and the abovementioned psychiatric conditions, it has been hypothesized that antipsychotic treatment could benefit "Internet addicted" patients. A promising case study reported the successful use of quetiapine (200 mg/day), gradually added to citalopram, in a 23 year old subject with IA (Atmaca 2007). The improvement was maintained at a 4-month follow-up.

10.2.5 Other Drugs

Future research on the pharmacological treatment of IA should also consider focusing on further drug categories, namely psychostimulants, alpha-2-adrenergic agonists (commonly referred to as alpha 2 agonists) and glutamatergic drugs.

Among these, psychostimulants (i.e. methylphenidate) are the only class for which anecdotal data is available in relation to a possible use in "internet addicted" patients. Methylphenidate (MPH) is a stimulant molecule, indicated for the treatment of attention deficit hyperactivity disorder (ADHD) in children and adults. It can also be used to treat chronic fatigue syndrome or symptoms of traumatic brain injury and daytime symptoms of fatigue induced by narcolepsy.

Recently, one trial tested methylphenidate in 62 attention-deficit/hyperactivity disorder (ADHD) subjects with "internet video game addiction", reporting a significant improvement both in attentional capacity and in Internet usage after 8 weeks of treatment (30.5 mg/day) (Han et al. 2009). The authors cautiously suggest that methylphenidate may be beneficial as a treatment for IA, especially where it co-occurs with ADHD. In fact, many clinical studies provided evidence for a link between ADHD and "Internet addiction", with comorbidity rates reaching up to 33 % (Yoo et al. 2004). This comorbidity suggests that other drugs used to treat ADHD patients could also be considered as avenues for possible treatment, at least for the subgroup of "Internet addicts" also suffering from ADHD. For instance, alpha 2 agonists, recently approved in controlled release formulations for ADHD, have anecdotally been found to act on impulsive behavior. In fact, among alpha 2 agonists, Guanfacine extended-release has demonstrated effectiveness in reducing impulsivity, hyperactivity and inattention in children and adolescents suffering from ADHD (Muir and Perry 2010). Likewise, there is evidence supporting the use of clonidine extended-release (another alpha 2 agonist) in the treatment of ADHD youth with inadequate response to stimulants. It appears that if clonidine is used in combination with psychostimulants, it provides incremental effectiveness in improving ADHD symptoms (Kollins et al. 2011). These preliminary clinical findings demonstrating that alpha 2 agonists have beneficial effects on ADHD symptoms that overlap with those of patients suffering from IA (namely impulsive behavior), suggest they should be studied in the context as IA as well.

Finally, the rationale for glutamatergic drugs' possible use in "Internet addicted" patients is linked to the fact that, along with dopaminergic dysfunction, glutamatergic system alterations have also been implicated in the pathophysiology of behavioral and substance addictions. There are several clinical reports that support the possible effectiveness of glutamatergic modulators in treating these conditions (Krystal et al. 2003). Among glutamatergic drugs, memantine (a NMDA receptor antagonist) and riluzole (an inhibitor of glutamate synaptic release) have been mostly investigated. Preclinical and clinical observations suggest that glutamatergic modulators target obsessive-compulsive symptoms and impulsivity. Memantine appears to diminish gambling and reduce impulsive decision making in patients with pathological gambling (Grant et al. 2010a). Of note, memantine was not more effective than placebo in reducing alcohol use (Evans et al. 2007). Riluzole has been found to have beneficial effects on patients with obsessive-compulsive disorder (Grant et al. 2010b) and in the treatment of self-injurious behavior associated with borderline personality disorder (Pittenger et al. 2005). Also, riluzole has been successfully used in cases of compulsive skin picking (Sasso et al. 2006) and in a patient with severe, chronic trichotillomania (Coric et al. 2007). Though promising, most of these results are reported in small studies and case studies and, thus, their generalizability is limited. Further examination of glutamate-modulating agents in the treatment of disorders associated with impulse control dysregulation and obsessive-compulsive symptoms, including IA, would certainly be of value.

10.3 Clinical Suggestions for a Psychopathologically Guided Approach

Case studies on IA treatment are rather limited, the quality of the current literature in this emerging field is not optimal, and no standard clinical treatment protocols or approved medications yet exist.

We summarize the reported evidence in Table 10.1.

The limited existing empirical evidence does not allow for definite conclusions to be drawn. We are not yet able to determine to what extent pharmacological treatment of approved psychopathological disorders may also help in the treatment of IA. But a clinically significant beneficial effect is easily discerned and has commonly been observed in daily clinical practice.

A recent meta-analysis of psychological and pharmacological interventions for IA suggests that both forms of therapy are highly effective for improving addictive behavior, time spent online, depression and anxiety after treatment (Winkler et al. 2013). With respect to pharmacological treatment, this analysis pooled 49 subjects of three different trials using escitalopram, bupropion and methylphenidate.

Table 10.1 Overview of clinical evidence on pharmacotherapy for IA

Class	Drug	Dosage	Patients	Outcome	References
Antidepressants	Escitalopram	30 mg/day for 3 months	1 Internet gaming addict	Mood improvement, reduction in online gaming drive and complete recovery of functioning	Sattar and Ramaswamy (2004)
		20 mg/day for 10 weeks	19 Internet addicts	Significant decrease in weekly hours spent online	Dell'Osso et al. (2008)
	Bupoprion	300 mg/day for 6 weeks	11 Internet gaming addicts	Decrease in craving and total time spent gaming online	Han et al. (2010)
		300 mg/day for 8 weeks	50 Patients with major depressive disorder and problematic online gaming	Reduction in online gaming and depressive symptoms	Han and Renshaw (2011)
Opioid receptor antagonists	Naltrexone	150 mg/day for >3 years (in addition to sertraline 100 mg/day)	1 Patient with compulsive cybersexual behaviour	Perceived control over sexual urges	Bostwick and Bucci (2008)
Antipsychotics	Quetiapine	200 mg/day for 4 months (in addition to citalopram 40 mg/day)	1 Patient with internet addiction	Lower obsessive-compulsive features and decrease in 'nonessential' internet use	Atmaca (2007)
Other	Methylphenidate	30.5 mg/day for 8 weeks	62 ADHD patients with video game addiction	Lower young internet addiction Scale scores and decrease in internet use	Han et al. (2009)

Interestingly, when comparing psychological versus pharmacological interventions, the authors did not find any significant differences in the efficacy of improving status and reducing time spent online. This finding supports the hypothesis that, in the future, pharmacotherapy could significantly contribute to the management of "Internet addicted" patients, with a particularly favorable cost-benefit profile.

In order to define the actual role of pharmacotherapy in the treatment of IA, we draw primarily on the evidence obtainable from clinical practice, to consider patterns of comorbidity and to propose some considerations from a psychopathological view. The latter are mere suggestions that may contribute to the complex clinical management of many "Internet addicted" patients, who are referred to clinicians, possibly alleviating their psychic distress and encouraging supporting their adhesion to treatment and rehabilitation programs.

If the patient exhibits high levels of discomfort and craving that interfere with the treatment strategy, and substance addiction coexists, the use of an opioid antagonist could be considered.

Patients with clinically significant anxiety levels or depressive symptoms, and specifically in case of comorbidity with anxiety disorders or depression, the use of a serotonergic drug could be helpful. In the case of a comorbidity with major depression, bupropion should be considered a good option. If the patient has manic or hypomanic symptoms, or in some cases of sub-syndromic excitement, the clinician could also consider the use of mood stabilizers, taking into account that their effectiveness in "internet addiction" has not yet been investigated.

To date, the research on IA is mainly focused on diagnostic criteria and assessment instruments (there is a significant need for consensus concerning clinical definitions and possible sub-forms relating to particular internet applications and/or activities). Future studies are needed to explore valid and reliable outcome measures and, certainly, further randomized controlled trials encompassing long-term follow-up data will be required to evaluate the treatment effects in large samples of "Internet addicted" patients. Research is also needed into whether addicts who use a particular medium/specific form of the Internet require different types of intervention.

References

American Psychiatric Association (2013) Diagnostic and statistical manual of mental disorders, 5th edn. American Psychiatric Association, Washington, DC

Atmaca M (2007) A case of problematic internet use successfully treated with an SSRI-antipsychotic combination. Prog Neuro-psychopharmacol Biol Psychiatry 31:961–962

Balodis IM, Kober H, Worhunsky PD et al (2012) Diminished frontostriatal activity during processing of monetary rewards and losses in pathological gambling. Biol Psychiatry 71:749–757

Bernardi S, Pallanti S (2009) Internet addiction: a descriptive clinical study focusing on comorbidities and dissociative symptoms. Compr Psychiatry 50:510–516

Bloch MH, Landeros-Weisenberger A, Kelmendi B et al (2006) A systematic review: antipsychotic augmentation with treatment refractory obsessive-compulsive disorder. Mol psychiatry 11:622–632
Blum K, Gardner E, Oscar-Berman M, Gold M (2012) "Liking" and "wanting" linked to reward deficiency syndrome (RDS): hypothesizing differential responsivity in brain reward circuitry. Curr Pharm Des 18:113–118
Bostwick JM, Bucci JA (2008) Internet sex addiction treated with naltrexone. Mayo Clin Proc 83:226–230
Bystritsky A, Ackerman DL, Rosen RM et al (2004) Augmentation of serotonin reuptake inhibitors in refractory obsessive-compulsive disorder using adjunctive olanzapine: a placebo-controlled trial. J Clin Psychiatry 65:565–568
Camardese G, De Risio L, Di Nicola M et al (2012) A role for pharmacotherapy in the treatment of "internet addiction". Clin Neuropharmacol 35:283–289
Choi YJ (2009) Efficacy of treatments for patients with obsessive-compulsive disorder: a systematic review. J Am Acad Nurse Pract 21:207–213
Christensen RC (2004) Olanzapine augmentation of fluoxetine in the treatment of pathological skin picking. Can J Psychiatry 49:788–789
Connor KM, Payne VM, Gadde KM et al (2005) The use of aripiprazole in obsessive-compulsive disorder: preliminary observations in 8 patients. J Clin Psychiatry 66:49–51
Coric V, Kelmendi B, Pittenger C et al (2007) Beneficial effects of the antiglutamatergic agent riluzole in a patient diagnosed with trichotillomania. J Clin Psychiatry 68:170
Davis RA (2001) A cognitive-behavioral model of pathological internet use. Comput Hum Behav 17:187–195
de Ruiter MB, Veltman DJ, Goudriaan AE et al (2009) Response perseveration and ventral prefrontal sensitivity to reward and punishment in male problem gamblers and smokers. Neuropsychopharmacology 34:1027–1038
Dell'Osso B, Hadley S, Allen A et al (2008) Escitalopram in the treatment of impulsive-compulsive internet usage disorder: an open-label trial followed by a double-blind discontinuation phase. J Clin Psychiatry 69:452–456
Denys D, de Geus F, van Megen HJ, Westenberg HG (2004) A double-blind, randomized, placebo-controlled trial of quetiapine addition in patients with obsessive-compulsive disorder refractory to serotonin reuptake inhibitors. J Clin Psychiatry 65:1040–1048
Di Chiara G, Bassareo V (2007) Reward system and addiction: what dopamine does and doesn't do. Curr Opin Pharmacol 7:69–76
Di Chiara G, North RA (1992) Neurobiology of opiate abuse. Trends Pharmacol Sci 13:185–193
Di Nicola M, Martinotti G, Mazza M et al (2010a) Quetiapine as add-on treatment for bipolar I disorder with comorbid compulsive buying and physical exercise addiction. Prog Neuropsychopharmacol Biol Psychiatry 34:713–714
Di Nicola M, Tedeschi D, Mazza M et al (2010b) Behavioural addictions in bipolar disorder patients: role of impulsivity and personality dimensions. J Affect Disord 125:82–88
Evans SM, Levin FR, Brooks DJ, Garawi F (2007) A pilot double-blind treatment trial of memantine for alcohol dependence. Alcohol Clin Exp Res 31:775–782
Garnis-Jones S, Collins S, Rosenthal D (2000) Treatment of self-mutilation with olanzapine. J Cutan Med Surg 4:161–163
Goddard AW, Shekhar A, Whiteman AF, McDougle CJ (2008) Serotoninergic mechanisms in the treatment of obsessive-compulsive disorder. Drug Discov Today 13:325–332
Goldberg JF, Whiteside JE (2002) The association between substance abuse and antidepressant-induced mania in bipolar disorder: a preliminary study. J Clin Psychiatry 63:791–795
Grant JE, Kim SW, Hollander E, Potenza MN (2008) Predicting response to opiate antagonists and placebo in the treatment of pathological gambling. Psychopharmacology 200:521–527
Grant JE, Kim SW, Odlaug BL (2009) A double-blind, placebo-controlled study of the opiate antagonist, naltrexone, in the treatment of kleptomania. Biol Psychiatry 65:600–606
Grant JE, Chamberlain SR, Odlaug BL et al (2010a) Memantine shows promise in reducing gambling severity and cognitive inflexibility in pathological gambling: a pilot study. Psychopharmacology 212:603–612

Grant P, Song JY, Swedo SE (2010b) Review of the use of the glutamate antagonist riluzole in psychiatric disorders and a description of recent use in childhood obsessive-compulsive disorder. J Child Adolesc Psychopharmacol 20:309–315
Griffiths M (2000) Does internet and computer "addiction" exist? Some case study evidence. CyberPsychol Behav 3:211–218
Guglielmo R, Martinotti G, Clerici M, Janiri L (2012) Pregabalin for alcohol dependence: a critical review of the literature. Adv Ther 29:947–957
Han DH, Lee YS, Na C et al (2009) The effect of methylphenidate on internet video game play in children with attention-deficit/hyperactivity disorder. Compr Psychiatry 50:251–256
Han DH, Hwang JW, Renshaw PF (2010) Bupropion sustained release treatment decreases craving for video games and cue-induced brain activity in patients with internet video game addiction. Exp Clin Psychopharmacol 18:297–304
Han DH, Renshaw PF (2011) Bupropion in the treatment of problematic online game play in patients with major depressive disorder. J Psychopharmacol 26:689–696
Hollander E, Rossi NB, Sood E, Pallanti S (2003) Risperidone augmentation in treatment-resistant obsessive-compulsive disorder: a double-blind, placebo-controlled study. Int J Neuropsychopharmacol 6:397–401
Hughes JR, Stead LF, Lancaster T (2007) Antidepressants for smoking cessation. Cochrane Database Syst Rev CD000031
Janiri L, Hadjichristos A, Buonanno A et al (1998) Adjuvant trazodone in the treatment of alcoholism: an open study. Alcohol Alcohol 33:362–365
Kim SW, Grant JE, Grosz RL (2002) Pathological gambling. Current status and new treatments. Minn Med 85:48–50
King DL, Delfabbro PH, Griffiths MD, Gradisar M (2011) Assessing clinical trials of Internet addiction treatment: a systematic review and CONSORT evaluation. Clin Psychol Rev 31:1110–1116
Ko CH, Liu GC, Hsiao S et al (2009) Brain activities associated with gaming urge of online gaming addiction. J Psychiatr Res 43:739–747
Kollins SH, Jain R, Brams M et al (2011) Clonidine extended-release tablets as add-on therapy to psychostimulants in children and adolescents with ADHD. Pediatrics 127:e1406–e1413
Koran LM, Bullock KD, Hartston HJ et al (2002) Citalopram treatment of compulsive shopping: an open-label study. J Clin Psychiatry 63:704–708
Koran LM, Aboujaoude EN, Gamel NN (2007) Escitalopram treatment of kleptomania: an open-label trial followed by double-blind discontinuation. J Clin Psychiatry 68:422–427
Krystal JH, Petrakis IL, Krupitsky E et al (2003) NMDA receptor antagonism and the ethanol intoxication signal: from alcoholism risk to pharmacotherapy. Ann NY Acad Sci 1003:176–184
Lee YS, Han DH, Yang KC et al (2008) Depression like characteristics of 5HTTLPR polymorphism and temperament in excessive internet users. J Affect Disord 109:165–169
Liu J, Gao XP, Osunde I et al (2010) Increased regional homogeneity in internet addiction disorder: a resting state functional magnetic resonance imaging study. Chin Med J 123:1904–1908
Maina G, Pessina E, Albert U, Bogetto F (2008) 8-week, single-blind, randomized trial comparing risperidone versus olanzapine augmentation of serotonin reuptake inhibitors in treatment-resistant obsessive-compulsive disorder. Eur Neuropsychopharmacol 18:364–372
Maremmani I, Pacini M, Lamanna F et al (2010) Mood stabilizers in the treatment of substance use disorders. Int J Neuropsychiatric Med 15:95–109
Martinotti G, Di Nicola M, Janiri L (2007) Efficacy and safety of aripiprazole in alcohol dependence. Am J Drug Alcohol Abuse 33:393–401
Martinotti G, Di Nicola M, Di Giannantonio M, Janiri L (2009) Aripiprazole in the treatment of patients with alcohol dependence: a double-blind, comparison trial vs. naltrexone. J Psychopharmacol 23:123–129
McElroy SL, Nelson EB, Welge JA et al (2008) Olanzapine in the treatment of pathological gambling: a negative randomized placebo-controlled trial. J Clin Psychiatry 69:433–440

Muir VJ, Perry CM (2010) Guanfacine extended-release: in attention deficit hyperactivity disorder. Drugs 70:1693–1702

Pallanti S, Quercioli L, Sood E, Hollander E (2002) Lithium and valproate treatment of pathological gambling: a randomized single-blind study. J Clin Psychiatry 63:559–564

Petry NM, O'Brien CP (2013) Internet gaming disorder and the DSM-5. Addiction 108:1186–1187

Pittenger C, Krystal JH, Coric V (2005) Initial evidence of the beneficial effects of glutamate-modulating agents in the treatment of self-injurious behavior associated with borderline personality disorder. J Clin Psychiatry 66:1492

Reuter J, Raedler T, Rose M, Hand I, Glascher J, Buchel C (2005) Pathological gambling is linked to reduced activation of the mesolimbic reward system. Nat Neurosci 8:147–148

Roncero C, Rodriguez-Urrutia A, Grau-Lopez L, Casas M (2009) Antiepileptic drugs in the control of the impulses disorders. Actas Esp Psiquiatr 37:205–212

Sasso DA, Kalanithi PS, Trueblood KV et al (2006) Beneficial effects of the glutamate-modulating agent riluzole on disordered eating and pathological skin-picking behaviors. J Clin Psychopharmacol 26:685–697

Sattar P, Ramaswamy S (2004) Internet gaming addiction. Can J Psychiatry 49:869–870

Shapira NA, Goldsmith TD, Keck PE et al (2000) Psychiatric features of individuals with problematic internet use. J Affect Disord 57:267–272

Shapira NA, Lessig MC, Goldsmith TD et al (2003) Problematic internet use: proposed classification and diagnostic criteria. Depress Anxiety 17:207–216

Shoja-Shafti S (2006) Treatment of borderline personality disorder with olanzapine. Arch Iran Med 9:403–405

Stewart RS, Nejtek VA (2003) An open-label, flexible-dose study of olanzapine in the treatment of trichotillomania. J Clin Psychiatry 64:49–52

Tao R, Huang X, Wang J et al (2010) Proposed diagnostic criteria for internet addiction. Addiction 105:556–564

Weinstein A, Greif J, Yemini Z et al (2010) Attenuation of cue-induced smoking urges and brain reward activity in smokers treated successfully with bupropion. J Psychopharmacol 24:829–838

Weinstein A, Lejoyeux M (2010) Internet addiction or excessive internet use. Am J Drug Alcohol Abuse 36:277–283

WHO (2010) The ICD-10 classification of mental and behavioral disorders, 10th edn. World Health Organization, Geneva

Winkler A, Dorsing B, Rief W et al (2013) Treatment of internet addiction: a meta-analysis. Clin Psychol Rev 33:317–329

Yen JY, Ko CH, Yen CF et al (2007) The comorbid psychiatric symptoms of Internet addiction: attention deficit and hyperactivity disorder (ADHD), depression, social phobia, and hostility. J Adolesc Health 41:93–98

Yoo HJ, Cho SC, Ha J et al (2004) Attention deficit hyperactivity symptoms and internet addiction. Psychiatry Clin Neurosci 58:487–494

Chapter 11
Therapeutic Interventions for Treatment of Adolescent Internet Addiction—Experiences from South Korea

Eunsuk Cho

Abstract This chapter introduces several intervention programs that have been developed and implemented for adolescents and younger Internet addicts. A few individual and group counseling programs currently operating in Korea will be outlined, and residential camps and integrative long-term therapy programs will also be introduced. The author also includes a summary of the characteristics of Internet-addicted youth in Korea.

The problem of Internet overuse has emerged in many countries over the past several years. With one of the most advanced IT infrastructures and almost universal access to the Internet, Korea has been particularly concerned with problems of Internet overuse since 2000. The Korean government and academia have made substantial efforts to prevent and treat the problem of Internet addiction. Many counseling and therapy services have been established, many of which have been government initiatives. Researchers and clinicians have worked together to develop and implement effective intervention models for those individuals experiencing Internet addiction.

The effects of Internet addiction on children and adolescents has been the source of much attention, due to the potential developmental implications. This chapter introduces many intervention programs that have been developed and implemented for adolescents and younger Internet addicts. The characteristics of Internet-addicted youth in Korea will briefly be reviewed, examples of individual and group counseling programs will be outlined, as well as residential camps and integrative long-term therapy programs which are now being operated in Korea.

Assistant professor of The University of Suwon. Previous director of the Myungji Internet Addiction Prevention and Intervention Center ("I Will Center") in Korea.

E. Cho (✉)
The University of Suwon, Hwaseong, South Korea
e-mail: successfny@gmail.com

C. Montag and M. Reuter (eds.), *Internet Addiction*, Studies in Neuroscience, Psychology and Behavioral Economics, DOI 10.1007/978-3-319-07242-5_11

11.1 Characteristics of Internet Addicts of Korean Adolescents

According to the annual national survey of Internet addiction (NIA[1] 2013), the rate of Internet addiction among Korean children and adolescents ranged between 9.4 and 11.7 % in 2012.[2] In other words, approximately 3–4 people per classroom require special monitoring of their Internet use. Work by Korean researchers and clinicians aimed at clarifying the characteristics of Internet addicts.

11.1.1 Internet Use Behaviors and the Observed Problems

According to NIA's 2012 survey (NIA 2013), on-line games were the most frequently used type of computer program by the addicted group, whereas mobile instant messenger applications were most frequently used among the non-addicted group. Daily average Internet use time for the Internet-addicted group was longer (3.2 h per day for the high risk group; 2.5 h for the potential risk group) than that of those in the normal group, which averaged 2.1 h per day. The majority of the problems arising from Internet addiction encompassed mental and physical health problems, parent-child relationship problems, and problems with school coursework.

Clinicians working in frontline centers observed that many Internet addicted adolescents faced problems such as failing to keep up with schoolwork, lying, poor peer relationships, stealing, being victims of cyber-crime, being tardy and coming home late, sleep deprivation, and irregular eating habits. Internet addicts are also likely to experience physical problems commonly related with computer overuse, such as eyestrain, neck pain, chronic fatigue, and weight gain. It is because of such problems that the clients and their families call for Internet addiction intervention services (Ahn 2012).

11.1.2 Psychological Traits

Irrational beliefs about Internet usage are unique characteristic of Internet addicts (Davis 2001). They think that the Internet will free them from stress and fatigue and connect them to other people and the world. This erroneous thinking is the focus of interventions using a cognitive behavioral approach (Young 2011).

---

[1] Korea National Information Society Agency (NIA) is one of the focal institutions of Korean government which was devoted to Internet addiction related research and the public response.

[2] Internet addiction was measured by KS Scale (Kim et al. 2008). % of high risk group: 1.8(elementary), 3.3(middle), 1.7(high school students) % of addicts (% of high risk and potential risk group): 9.4(elementary), 11.7(middle), 9.6(high school students).

High impulsiveness and low self-control are the most influential psychological traits of Internet addicts. Because of these aspects, Internet addiction maybe considered as a form of Impulse-Control Disorder (Lee et al. 2012). Comorbid symptoms such as ADHD (Yen et al. 2007) can be understood in the same context. For this reason, clinicians frequently include the learning of self-regulation skills in their approach to Internet addiction counseling.

Internet addicts' lack of problem-solving skills often leads them to unhealthy coping styles such as "avoidance". Instead of confronting and coping with difficulties, they cling to the Internet to avoid their problems, thereby adding further to such problems (Whang et al. 2003). Stress management and coping techniques have been incorporated into counseling protocols to help address these issues.

In many cases, Internet addicts experience depression as a comorbid symptom (Ha et al. 2007; Kim et al. 2006). They feel isolated and their self-esteem appears to be lower on average compared to non-addicts (Kim and Davis 2009). They feel safe in the Internet world where anonymity seems to be guaranteed. For these depressive clients, finding and dealing the exact causes of their depression is necessary, in addition to treating their symptoms of IA.

11.1.3 Family and Peer Relationship

Oftentimes, parents of Internet addicted adolescents struggle to develop the proper parenting skills to deal with the issues their children face. Extreme permissiveness (i.e. through a lack of rules and boundaries) or extreme coercive control over their children's behavior becomes an important factor in their children's Internet overuse (Song and Park 2008). When this is the case, psychological education and counseling for the parents is also important and necessary.

Lack of family resources is another characteristic of Internet addicts. This is observed in children who have "nothing to do after school" and resort to the Internet. Parents of children suffering from Internet addiction commonly work long hours, coming home late in the evenings, and have no extra resources to supplement parental care at home (Choi et al. 2009). In such instances, it is helpful to connect the family to community resources, provide a mentoring service to the children, and organize afterschool programs as this provides a safe, Internet-free environment.

In some cases, it is observed that there are serious family conflicts such as parental separation, divorce or violence. In other cases, there may be severe conflict in the parent-child relationship due to Internet use problems (Choi et al. 2009; Song and Park 2008). In both cases, family counseling is suggested for disentangling family issues.

In many cases, addicted adolescents do not have satisfying peer relationships. They have a tendency to withdraw from their peer groups and isolate themselves from the real world (Bae et al. 2012c). This is sometimes caused by a lack of social skills. Social skills training and group counseling are effective in promoting positive peer interactions for such clients.

11.1.4 Implications of Counseling Interventions

We suggest several implications for counseling interventions from the above review of the characteristics of Internet addicted adolescents. First, an accurate initial assessment of the addicts is necessary for effective intervention. While the manifest behavior of addictive use of the Internet may seem similar, the addiction arises from various roots (Douglas et al. 2008) including temperamental, cognitive, psychological and familial factors. Also, comorbid symptoms need to be screened for proper psychiatric treatment and counseling. For the initial assessment, the counseling office needs to prepare an interview sheet to check the client's Internet use behavior, addiction level, and psychological assessment tools.

Second, as with other forms of addiction counseling, the counselor must first approach the client using motivational interviewing techniques (Miller and Rollnick 2002). This is particularly important when the client is a child or an adolescent, as their participation in counseling is typically involuntary and they often prefer to ignore their problem. Showing genuine interest in the client's life (including the client's favorite Internet game) is important. It is also important to foster good rapport in group-counseling programs, e.g. through the provision of snacks and the facilitation of interesting activities.

Third, counseling needs to tackle the immediate and practical issues first and deal with deeper concerns later (Bae et al. 2012a). Many Internet addicts who come in for treatment suffer from various life problems, such as failure in school, irregular eating and sleeping patterns, etc. Counselors should first facilitate a return to everyday basics in the client's life, before approaching behavioral and cognitive issues like time management, planning, communication skills, study methods, irrational beliefs, stress coping skills and other activities. If there are family conflicts or issues with the client's inner-self, the counselors can then go further and deeper to help resolve these issues.

Finally, relapse management is important. Preparing skills to cope with relapse and risky situations for the clients and their family is an essential goal in the final phase of counseling. Mentoring services, club activities and support groups can be provided to the clients or their families after the counseling or intervention has finished.

11.2 Individual Counseling

11.2.1 Counseling Service System

This section introduces the Internet addiction counseling service system for children and adolescents in Korea. Korea has implemented a thorough nationwide screening system, which in turn has increased demand for more intervention services and more research to be done in the field of Internet addiction.

Enumeration survey of Internet addiction: Since 2009, Korean government has implemented an annual enumeration survey of Internet addiction for the target grades of 4, 7 and 10. This is a very thorough screening system designed to detect those at risk of developing an Internet addiction. Based upon the survey results, counselors select students in the high-risk group and then assess their comorbid symptoms using Child Behavior Check List (CBCL by parental report for grade 4) or Youth Self Report (YSR for grade 7 and 10) (Achenbach and Rescorla 2001). They refer those students who show clinical psychiatric symptoms to hospitals for more professional psychiatric assessment and treatments. For high-risk students who do not have comorbid symptoms, they provide individual or group counseling service throughout the year. The counseling service is mostly offered at the school by the school counselors or by counselors from local counseling centers.

Central dissemination of the expertise of the Internet addiction counseling: To meet the vast amount of counseling requests after the survey, the government offered professional training in Internet addiction counseling through the government institution, the Korea Youth Counseling and Welfare Institute (KYCI), and the National Information Society Agency (NIA). These two institutions are developing individual counseling protocols and group counseling programs for each age group, and disseminate the program manual for frontline counselors to use and to improve the quality of services.

Home Visit Counseling service: Home Visit Counseling (HVC) services, in which counselors visit clients in their own homes, are provided for clients who are socially-withdrawn or disconnected from social services due to disadvantaged home environments. The visiting counselors offer the initial assessment and initiate a short-term counseling intervention (i.e. 3–4 sessions). Actually, the goal of HVC services is not to complete the treatment in the home setting, but to motivate these clients come to the counseling office for further counseling and treatment (Choi et al. 2009).

11.2.2 Counseling Protocol

To guide the quality of Internet addiction counseling in the local youth counseling centers, KYCI initiated the development of an individual counseling protocol (Bae et al. 2012c). This protocol is now disseminated through the regular in-depth counselor training offered by KYCI.

For the more effective interventions, the researchers (Bae et al. 2012c) developed different protocols based on subtypes of the addicts. For this purpose, they categorized Internet addicts into three groups according to clients' temperament, affect, and relationship dimensions: "the stimulus seeking type", "the depressive type", and "the weak peer relationship type". A different protocol was designed for each type to effectively target the individual root causes of addiction, which is likely to differ between individuals, despite similar surface presentation of symptoms. The protocols are summarized in Table 11.1.

Based on experimental trials of the protocol in 12 cases, the intra-group change indicated this counseling protocol helped to decrease the duration of Internet use

Table 11.1 KYCI's Internet addiction individual counseling protocol (Bae et al. 2012c)

Internet addiction counseling protocol (Bae et al. 2012c)	
• **Adolescent counseling**	
Initial phase (Session 1–3)	Building rapport, assessment, motivational interview
	Recognizing the problem, goal setting
Middle phase (Session 4–10)	*Stimulus seeking type*: self-regulation training
	Depressive type: CBT approach to treat depression
	Weak peer relationship type: social skills training
	Common for all types: changing irrational thoughts on Internet use, time planning, finding alternative activities to replace the internet, coping skills, career plans, etc.
Final phase (Session 11–12)	Creating a supportive environment, planning for relapse, evaluation
• **Parent counseling**	
Initial phase (Session 1–2)	Building rapport, motivational interview
	Parental assessment, sharing the child's assessment results with the parents
	Therapeutic goal setting for the parent and the adolescent
Middle phase (Session 3–5)	*Stimulus seeking type*: supportive parenting for the child's self-regulation
	Depressive type: encouraging and supporting the depressive child
	Weak peer relationship type: supportive parenting for the child's better peer relationship
	Common for all types: Dealing with parent-child conflict, Parenting skills in guiding Internet use for the child
Final phase (Session 6)	Planning for relapse, evaluation

and the level of Internet addiction of the clients. Additionally, positive changes were observed in peer and in parent-child relationships, self-regulation and depressive symptoms of the clients, and/or the parent(s) (Bae et al. 2012c).

11.2.3 Hospital Treatment Model

Compared to psychological counseling services, hospital services are often specialized in accurate diagnosis of the psychiatric symptoms and proper medications to relieve the symptoms. Hospitalization is, sometimes, the last choice for very serious addiction clients.

One hospital-run center is the "On-Line Game Clinic and Research Center", operated by Chung-Ang University Hospital. In addition to regular psychiatric treatment, they provide individual and group counseling, family counseling and sports therapy in an integrative mode for Internet addiction patients. Known for their active neuro-scientific research on Internet addiction, the "On-Line Game Clinic and Research Center" is continuously developing effective treatment methods.

A recent publication highlighted the efficacy of CBT individual therapy combined with medication (bupropion) for treatment of problematic on-line game play (Kim et al. 2012).[3] The clinic has also developed an applied sports therapy treatment, which has proven an effective method in improving attention, cognitive symptoms and social skills in ADHD children (Kang et al. 2011), and Internet addicts, who suffer from similar symptoms.

11.3 Group Counseling

Group counseling has a unique effect on Internet addicted adolescents in two ways. First, the counseling group can be both a reference group and a support group for the individual client. Their unstable and fragile self-identity needs and wants approval from peer-groups. A perceived lack of peer approval is one of the main contributors to adolescents' Internet (gaming) addiction and is often the underlying reason for their resistance to quit Internet overuse. Peers in counseling groups provide healthy peer pressure and approval, facilitating change in addicts' lifestyles. In this way, they work as a support group to help each other overcome their Internet addiction. When one or two leading figures change and overcome their addictions, it creates a large positive impact on the other resistant group members, an effect that cannot be achieved through individual counseling.

Second, the clients are able to build relationships in group-counseling. Many clients have not been satisfied with their peer relationships, partly because of their poor social skills. The safe and accepting atmosphere of the group encourages group members to be active and to have social interactions with other members, which, in turn, creates positive relationship experiences for them. For these reasons, group counseling is a popular treatment for children and adolescents.

Group counseling programs are implemented in schools following administration of the enumeration survey, and are facilitated by counselors dispatched from local counseling centers. To motivate client participation, most of the programs have activities such as games, drawing, crafts, and cooking, in addition to psychoeducation about Internet addiction interventions. They aim to facilitate "learning through activities" while taking the clients' developmental characteristics into consideration.

According to Park (2009)'s meta-analysis of 41 Internet addiction group counseling programs in Korea, the main theoretical approaches of the group counseling programs are "integrative" (mixture of effective interventions from different approaches), cognitive-behavioral therapy (CBT), motivational enhancement therapy (MET: Miller 1999), reality therapy (Kim 2008), solution-focused therapy

[3] The 'CBT & medication' group showed the reduction of internet addiction score more compared to the 'medication only' group (Kim et al. 2012).

Table 11.2 Example of Internet game addiction group counseling program for middle school students (Hyun et al. 2006)

Session	Goal		Stage
1	• Building rapport	• Exploring the pros and the cons of the change	Precontemplation
	• Recognizing the risks of game addiction	• Motivating for change	
2	• Recognizing the underlying motives of Internet game use	• Weighing the pros and the cons of the Internet game and the change	Contemplation
	• Dealing with the ambivalence toward change		
3	• Finding my strength	• Finding alternative activities	Contemplation
	• Reflecting my successful experience	• Exploring the benefits of the change	
4	• Understanding the parents' mind	• Parental pledge of supporting the child's alternative activities	Preparation
	• A pledge of abstinence (control) of Internet game		
5	• Finding the reasons of clinging to the Internet game	• Finding the coping strategies to overcome the temptation of Internet game	Action
6	• Checking the high risk situation related to peer pressure	• Role play of saying "No"	Action
	• How to say "No" to the temptation		
7	• Exploring my dream	• Consolidate the dream to bridge it to reality	Action
8	• Sharing my change	• Tamping my resolution	Maintenance
	• Coping skills training for relapse and risky situation	• Evaluation	

(Moon et al. 2011), and expressive art therapy (Chung 2008). Park (2009) concludes that Internet addiction group counseling programs in Korea are effective in reducing addiction level. Of these therapies, the integrative approach, reality therapy, and CBT show the greatest efficacy.[4]

One example of a group-counseling program is presented in Table 11.2. This program is based on a combination of the trans-theoretical model (Prochaska and DiClemente 2005) and the CBT model. It is targeted for Internet game addicted middle school students (7–8 participants). The post-test result indicates this program is effective in significantly reducing Internet addiction scores (Hyun et al. 2006).

[4] Park (2009) evaluated the 195 effect sizes of the 41 group counseling programs and got 1.04 mean effect size which means quite high effectiveness.

11.4 Residential Therapy Program

Residential therapy programs such as intensive "school" or "camp" to help clients recover from Internet addiction are also popular in Korea. A variety of programs are available according to the clients' situations, such as "one day family camp", "daycamp (for 2–3 days)", "12 day intensive treatment", or "long-term (a couple of months) residential treatment"[5]. The content of residential programs comprise individual and group counseling, sports or outdoor activities, artistic or play activities, family activities, and parental education. Regular eating, sleeping, walking, social interaction and full abstinence from the Internet are the "unique" benefits of the residential program. In fact, these "regular" daily life practices play important role in the recovery of Internet addicts.

KYCI's "Internet RESCUE School" is known as the most successful residential program in Korea (Koo et al. 2011). It has been operated and modified since 2007, and the operation manual was published in 2010 (Hwang et al. 2010). Now, this 12-day residential program is disseminated by KYCI's special workshop for local treatment centers and is expanding rapidly beyond metropolitan areas to the entire country with the support of the government. The target group of this "RESCUE School" is high-risk Internet addicted adolescents. The goals of this program are reducing (controlling) Internet use, improving parent-child relationship, learning stress coping strategies and career planning of the participants. It employs and combines the ecological perspective integrating bio-psycho-social aspects, Motivational Enhancement Therapy, behavioral therapy, CBT and solution-focused approach. The summary of the program is presented in Table 11.3.

A 2–3 months' screening process is implemented before the main program starts. Clients with serious comorbid symptoms that would hinder program participation or trouble group dynamics are screened out. Three parental visits to the "School" during the 12 days are mandatory. Because the program is usually offered in the countryside surrounded by an abundance of natural environments and facilities for youth activities, it is hard for some families to come to the "School" during weekdays. Throughout the "School" days, individual and group counseling are provided. Parental counseling and family therapy are also provided during family visitation days. A regular day at the "School" starts with a walk in the forest and closes with meditation. Sports activities such as soccer are enjoyed during the day. Therapeutic activities such as group projects, energizing activities and artistic works are also provided. There is also a "Two-day family camping" program at a tent site which consists of family recreation, cooking and experiences that facilitate positive interactions among family members.

A mentor is matched to each participant during the whole 12 days. Mentors support and encourage the clients' successful participation in the program. At the beginning of the "School", most clients are very resistant and terrified by the fact

[5] For now, there is only one residential treatment center in Korea, which is funded by a Christian church. Another one will be opened soon, which is operated by government.

Table 11.3 The summary of the "RESCUE school" of KYCI (Hwang et al. 2010)

Screening	• **Inclusion criteria**:
	- Internet addiction high risk group (KS-scale[a] & Intake interview)
	• **Exclusion criteria:**
	- Serious psychiatric symptoms hindering the "School" participation
	- Parents cannot participate in the three parental sessions (for 5 days) during the "School" days
Phase I	• Group counseling: sessions 1–3
	• Individual counseling: "recognizing my addiction"
	• Parental education: understanding the child's psycho-emotive status
	• Therapeutic activities for relationship building
	• Other therapeutic activities: strolling, meditation, sports
Phase II	• Group counseling: sessions 4–6
	• Individual counseling: solution-focused approach to the issues related to Internet addiction
	• Parental intervention: family camping, parent counseling, family therapy
	• Therapeutic activities for energizing the clients: wall-climbing, skin-scuba diving, etc.
	• Other therapeutic activities: Meaningful experiences(i.e., volunteering at the neighborhood) added to the above activities
Phase III	• Group counseling: session 7
	• Individual counseling: family relationship, coping strategies
	• Parental intervention: "recognizing the change of the child"
	• Therapeutic activities for recovery of positivity: pottery, craft, etc.
	• Other therapeutic activities: strolling, meditation, sports
Follow-up	Parental education and support group
	Mentoring for sustaining the change (1:1 mentoring service)
	Alternative activities (sport, hobby, etc.)

[a]KS Scale is an Internet addiction scale developed by Kim et al. (2008). This is the most popular measurement of Internet addiction in Korea, used for national survey as well as many Internet addiction counseling assessment

that they will be disconnected from the Internet. However, they become gradually accustomed to "the School life" and build up their new lifestyle. Of course, a few members fail to overcome their addiction and leave the "School" during the program. After graduation from the "School", participants undertake three-months of follow-up programs, such as mentoring services and regular parental education sessions to help parents sustain the change in their children.

The intra-group changes show that participants' Internet-usage times significantly decrease and show positive changes in Internet-use-related aspects of the adolescents and their parents (e.g. thoughts about using the Internet, career development, parenting style, parent-child communication, wellbeing, self-control). Furthermore, the change was sustained at a two-month follow-up after the program finished (Hwang and Du 2011).

Because KYCI's "Internet RESCUE School" is available only to adolescents whose parents can cooperate, an alternative residential camp was developed by

KYCI in 2012 for adolescents whose parents cannot visit during the camp. For elementary school students who have problems with usage of the Internet, two or three day family camp programs are usually offered. In these programs, the focus is more on parental education and parent-child communication (Bae et al. 2012b).

11.5 Long-term Therapy Program

Long-term integrative programs have been implemented by "I WILL" centers.[6] These programs were developed in answer to the question, "Why do adolescents who seem to have overcome addiction relapse so easily?" They found that the origin of the problem was the "fragile self" of the clients. And thus they designed a program in order to tackle this problem. The program aims to build up a balanced, healthy life through a year-long period of therapy. The program focuses not just on reducing Internet usage, but also on strengthening the "root" of the addicts; this includes things such as their self-image, self-confidence, and future dreams. "Gwang-Jin I WILL Center" began the first long-term program in 2009. Through several iterations, this program has come to have its present shape, displayed in Table 11.4. The 4-1-4 months' module of the main program and the one-year follow-up is designed to fit the Korean school calendar.

The program is intended for high-risk Internet addicted middle school students, upon recommendation of school teachers. Priority is given to clients from disadvantaged home environments. About 15 members are included in each cohort. In many cases, participants are found deprived of various cultural experiences, sports activities and achievements in their life. Internet gaming is their only respite and amusement. Phase I consists of mostly counseling programs. Once it finishes, Phases II and III are focused on providing various experiences (e.g. music, acting and performing, outdoor camping, family activities, etc.) to offer opportunities "to learn through activities". A follow-up program is offered to maintain the change and to facilitate further growth through mentoring and club activities (e.g. sports, hobbies, etc.).

Because of high costs and other operational difficulties, such as retaining the participants during the 1+ year duration of the program, practitioners debate the long-term efficacy of this program versus its cost-benefit ratio (Cho and Kim 2011). However, the impressive results observed in clients—even several years after completing the program—and the relationship experiences between clients and group leaders, motivate practitioners to contribute their passion to this program. According to the program evaluation summary (Ahn 2012), the changes in Internet addiction scores of the participants reflect the effectiveness of this program. Qualitative materials such as letters from school teachers and families, or

[6] "I WILL Center" is the Internet addiction prevention and intervention center run by Seoul Metropolitan Government, Korea.

Table 11.4 Long-term integrative program "Dream Tree" of Gwang-Jin I WILL Center (Ahn 2012)

Long term integrative therapy program "Dream Tree"				
	Title	Client's status	Goal	Therapy/activities
Phase I (4 months)	Recovery of my "self"	• Psychological weaknesses	• Recognizing the problem and starting to change Internet use habit	• Psychological assessment
				• Individual counseling
		• Lack of self-control		• Group counseling
				• Parent counseling
		• Family conflicts		• Family activity
Phase II (1 month)	Exploring myself	• Lacks diverse experiences	• Discovering my interest	• Outdoor camping
			• Experiencing joy	• Cultural activities
		• Lacks ability to deal with stress		• Group counseling
		• Passive attitude		
		• Lethargic		
Phase III (4 months)	Self-expression	• Lacks success experience	• Experiencing a sense of achievement	• Alternative activities
				• Performance/ exhibition
		• Lacks self- confidence		• Graduation ceremony
		• Lacks life satisfaction		
Phase IV (12 months)	Follow-up maintenance	• Relapse possibility	• Maintaining self-control and the changed habit of Internet usage	• 1:1 mentoring
		• Limited resources		• Youth self-help group, (club) activities
		• Lacks initiative and future plan		

participants' evaluations show a profound change in the participants' lifestyle. Further efforts are needed to modify and to develop the operation manual of this program as well as to provide scientific research on the effectiveness of this program.

11.6 Conclusion

In the midst of its IT prosperity, Korea is suffering from the affliction of Internet addiction. However, with the active response of the Korean government, scholars and practitioners have continuously developed and implemented better treatments for Internet addiction. From Korea's various intervention experiences, we can draw several implications for the future of Internet addiction treatment for children and adolescents.

First, Internet addiction treatment for children and adolescents cannot be separated from parental or family counseling. Parents sometimes cannot play their parental role of supervising and guiding their children's Internet usage for various reasons. It can be due to serious marital conflict, economic difficulties, lack of time, or lack of effective parenting skills. Whatever the reason, tackling the risk factors on the parental side is one important key to resolving the child's addiction problem. Consequentially, a substantial number of counseling programs in Korea use the multi-modal approach, providing treatment for children/adolescents as well as parental/family counseling in parallel (Bae et al. 2012c; Hwang et al. 2010; Song and Park 2008).

Second, clinicians and scholars need to pay more attention to non-traditional counseling methods in treating children and adolescents' Internet addiction problems. That is, the effect of positive experiences and therapeutic activities in the treatment of the addiction cannot be ignored. Of course, the need for counseling and psychiatric treatment is indisputable. However, the experience of being together with peers, being deeply involved in other hobbies, and developing emotional closeness with their parents brings out their potential growth. For better treatment of children and adolescents' Internet addiction, more creative intervention programs are expected in the future that integrate positive experiences, therapeutic activities as well as counseling services.

Third, clinicians need to have close relationships with the clients' school teachers and parents. This is crucial in detecting addictive problems early and treating them successfully. We cannot expect voluntary participation of younger clients in treatment, especially during the early stage of the program. In fact, teachers and parents play key roles in maintaining the treatment of younger clients. For this reason, in Korea, many educational workshops are offered to parents and teachers.

Finally, we see Internet overuse as the beginning signs of an addiction that needs to be treated in the early stage. If interventions do not occur properly and in a timely manner, it gradually undermines the daily life of affected children and adolescents up to an irrevocable level. Therefore, the importance of the regular assessment of Internet addiction and the preventive education about the risk of the addiction for every children and youths needs to be addressed.

References

Achenbach TM, Rescorla LA (2001) Manual for the ASEBA school-age forms and profiles. ASEBA

Ahn YH (2012) A counseling approach to youth Internet addiction: Seoul Metropolitan Government "I WILL Center" counseling program. In: Proceedings of 2012 Asian youth workers training workshop. Seoul Youth Center for Cultural Exchange, Seoul, Korea

Bae JM, Cho YM, Kim KE (2012a) Counselors' perspectives on the characteristics of Internet-addicted youth counseling. Korean J Couns 13(6):2733–2761

Bae JM, Cho YM, Chung HY (2012b) Development of family healing camp program for the Internet addicted adolescents. Korea Youth Counseling & Welfare Institute Research Report

Bae JM, Yang JW, Cho YM, Kim KE, Choi YH (2012c) Development of an individual counseling protocol for Internet-addicted youths. Korea Youth Counseling & Welfare Institute Research Report

Cho ES, Kim HJ (2011) A model of long-term therapy program for Internet-addicted adolescents. In: Proceedings of Korean Addiction Counseling Association 2011 annual fall conference, pp 121–131

Choi DJ, Kim DM, Koh YS, Cho YM, Lee EA, Lee ES (2009) A study on developing a guideline of an outreach program of Internet–addicts. National Information Society Agency Research Report

Chung YJ (2008) Effects of integrative arts therapy on Internet use control and self-control of Internet addictive high school students. Korean J Art Ther 15(3):555–570

Davis RA (2001) A cognitive-behavioral model of pathological internet use (PIU). Comput Hum Behav 17(2):187–195

Douglas AC, Mills JE, Niang M, Stepchenkova S, Byun S, Ruffini C, Lee SK, Loutfi J, Lee JK, Atallah M, Blanton MA (2008) Internet addiction: meta-synthesis of qualitative research of the decade 1996–2006. Comput Hum Behav 24:3027–3044

Ha JH, Kim SY, Bae CS, Bae S, Kim H, Sim M, Lyoo IK, Cho SC (2007) Depression and Internet addiction in adolescents. Psychopathology 40:424–430

Hyun MH, Chang JH, Chung TY, Joo HW, Jeon HY, Kang MJ (2006) Development of counseling program for Internet-game overuse adolescents. Korea Game Industry Development Agency Research Report

Hwang SG, Lee HC, Cho KP, Oh HY, Du JY, Lee YH (2010) Development of the residential treatment program for internet addicted adolescents. Korea Youth Counseling Institute Research Report

Hwang SG, Du JY (2011) The effectiveness evaluation of the residential treatment program for internet addicted adolescents. Ment Health Soc Work 37:145–188

Kang KD, Choi JW, Kang SG, Han DH (2011) Sports therapy for attention, cognitions and sociality. Int J Sports Med 32:953–959

Kim SM, Han DH, Lee YS, Renshaw PF (2012) Combined cognitive behavioral therapy and bupropion for the treatment of problematic on-line game play in adolescents with major depressive disorder. Comput Hum Behav 28(5):1954–1959

Kim K, Ryu E, Chon M, Yeun E, Choi S, Seo J, Nam B (2006) Internet addiction in Korean adolescents and its relation to depression and suicidal ideation: a questionnaire survey. Int J Nurs Stud 43(2):185–192

Lee HW, Choi J, Shin Y, Lee J, Jung H, Kwon J (2012) Impulsivity in Internet addiction: a comparison with pathological gambling. Cyberpsychol Behav Soc Network 15(7):373–377

Kim DI, Chung YJ, Lee EA, Kim DM, Cho YM (2008) Development of Internet addiction proneness scale-short form (KS scale). Korean J Couns 9:1703–1722

Kim H, Davis KE (2009) Toward a comprehensive theory of problematic internet use: evaluating the role of self-esteem, anxiety, flow, and the self-rated importance of Internet activities. Comput Hum Behav 25(2):490–500

Kim JU (2008) The effect of a R/T group counseling program on the internet addiction level and self-esteem of internet addiction university students. Int J Reality Ther 27(2):4–12

Koo C, Wati Y, Lee CC, Ho HY (2011) Internet-addicted kids and South Korean government efforts: boot-camp case. Cyberpsychology Behav Soc Networking 14(6):391–394

Korea National Information Society Agency (NIA) (2013) 2012 National survey of Internet addiction

Miller WR (ed) (1999) Enhancing motivation for change in substance abuse treatment, vol 35. DIANE Publishing

Miller WR, Rollnick S (2002) Motivational interviewing: preparing people for change. Guilford Press, New York

Moon HS, Kim OH, Koh YS, Bae SM (2011) Effectiveness of solution focused group counseling for adolescents with Internet addiction. Korean J Fam Ther 19(3):123–140

Park SM (2009) A meta-analysis on the effects of adolescent internet addiction group counseling program in Korea. Korean J Couns Psychother 21:607–623

Prochaska JO, DiClemente CC (2005) The transtheoretical approach. In: Norcross JC, Goldfried MR (eds) Handbook of psychotherapy integration. Oxford University Press, New York, pp 147–171

Song SM, Park SM (2008) Development of Internet addiction intervention program according to the subtype of the addicts. Korean Youth Counseling Institution Research Report

Whang LS, Lee S, Chang G (2003) Internet over-users' psychological profiles: a behavior sampling analysis on internet addiction. CyberPsychology Behav 6(2):143–150

Yen J, Ko C, Yen C, Wu H, Yang M (2007) The comorbid psychiatric symptoms of internet addiction: attention deficit and hyperactivity disorder (ADHD), depression, social phobia, and hostility. J Adolesc Health 41(1):93–98

Young KS (2011) CBT-IA: the first treatment model for Internet addiction. J Cogn Psychother Int Quart 25(4):304–312

Chapter 12
Therapeutic Interventions in the Treatment of Problematic Internet Use—Experiences from Germany

Wolfgang Dau, J.D.G. Hoffmann and Markus Banger

A journey of a thousand sites begins with a single click
(*Author Unknown*).

Abstract In this chapter, the Internet usage behavior of the general population and the prevalence of problematic Internet use will be discussed and a variety of treatment options are introduced. Theoretically and practically, through use of a case study, the brief intervention, "Compass", is described. The last section considers research on the efficacy of existing approaches to the treatment of problematic Internet use. To contextualize the prevalence of Internet use, the underlying concepts and the therapy, this chapter opens with a brief consideration of television and the debate about TV dependency, which took place during the 1990s.

Before taking a closer look at the prevalence of Internet use, the underlying concepts and the therapy, this chapter begins with a brief recourse on television and the debate about TV dependency, which took place during the nineties of the last century. For the triumph of the Internet will possibly go hand in hand with the "death" of another medium: television, in any case, in its "classical" form (Katz and Scannell 2009). At least with youngsters, television seems to have become less important compared with the Internet. For instance, among 12–19 year olds, as far as frequency of media activity went, television came in third place in 2011, after the mobile phone and the Internet. Among boys, 89 % considered personal use of the Internet very important/important, while only 58 % made this statement about television. Among girls, the Internet came after "listening to music" and "mobile phone use", in third place with 86 %, and television was in sixth place with 54 % (Medienpädagogischer Forschungsverbund Südwest (MPFS) 2012). In 2009, Internet and television were of equal importance to this age group; television

W. Dau (✉) · J.D.G. Hoffmann · M. Banger
Department for Addiction and Psychotherapy, LVR-Clinic Bonn, Bonn, Germany
e-mail: Wolfgang.Dau@lvr.de

C. Montag and M. Reuter (eds.), *Internet Addiction*, Studies in Neuroscience, Psychology and Behavioral Economics, DOI 10.1007/978-3-319-07242-5_12

may even have had a slight edge (MPFS 2010). A note on terminology is necessary here. A key problem in looking at the phenomenon of Internet use is the veritable flood of terminology—much of which may also be used differentially; the concept "problematic Internet use (PIG)" will be consistently used in this chapter, except in the reproduction of results from studies that explicitly use a different terminology.

12.1 Media Socialization and Moral Panic: "Digital Immigrants" and "Digital Natives"

Hurrelmann and Quenzel (2012) refer to adolescents who naturally grew up with digital new media, as "digital natives", while the parents, being representative of the older generation, can be seen as "digital immigrants". In the wake of World War 2, television began its slow but inexorable ascent in popularity, which socialized the generation of "digital immigrants" in media terms. Consequently, this generation is characterized and/or is influenced by the earlier discussions about the negative consequences of television use (Livingstone 2009). The resurgent discussion, particularly in Germany, that accompanied the introduction of private television in 1984 is still in the memory of the "digital immigrant" generation and seems to assume certain aspects of 'moral panic'. A moral panic is an intense feeling expressed within a population over an issue that appears to threaten the social order (Jones and Jones 1999). A well-known media critic primarily associated with television, was the communication scientist Neil Postman who made the criticism "television is in the process of transforming our culture into a gigantic arena for show business" (Postman 1988, p. 102). In this context (Livingstone 2009) pointed out that

> [the] moral panics associated with the arrival of each new medium, which demand that research address the same questions over and again – about displacement of reading, exercise, and conversation; about social isolation and addiction; about violent and consumerist content [...]–have a long history (p. 152).

Accordingly, the discussion about television dependence that occurred during the nineties closely resembles to the debate on Internet dependence. At that time researchers had already begun to debate the definition and/or existence of substance-related and non substance-related addictions, about possible addiction-generating features of the medium, specific characteristics of those people susceptible to addiction and over the difficulties regarding exact diagnostic criteria, as well as problems of comorbidity (Horvath 2004; McIlWaraith et al. 1991). TV dependence has so far not been included in diagnostic manuals and whether inclusion will be extended to the question of Internet dependence remains to be seen.

12.1.1 Compulsion? Addiction? Impulse Control Disorder? Search for a Usable Definition for Therapy

Attempts to define "Internet addiction" or "media dependence" present particular difficulties in that, very often, multiple phenomena may be included in these terms, e.g. the excessive use of pornographic content, games, general surfing of the Internet and excessive use of social networks (Winkler et al. 2013). It has been recently pointed out, in fact, that separate forms of generalized and specific Internet addiction must be distinguished (Montag et al., accepted).

Nevertheless, clear definitions and concepts of disorders are very important for the planning of treatment and therapy. Furthermore, it is not always clear whether the individuals concerned are dependent *in* or *on* the Internet (Griffiths 2000). Classification of the research findings to date has been further complicated by the absence of uniform diagnostic criteria and measurement instruments (Winkler et al. 2013). At the moment, such classification is largely dependent on the criteria of substance-related addictions or pathological gambling (Rehbein et al. 2013). At least for the so-called "Internet Gaming Disorder", the American Psychiatric Association has decided to include nine criteria (see Table 12.1) in the Appendix of DSM-V as a research diagnosis (American Psychiatric Association 2013). These criteria are not intended for clinical use, but to encourage future research.

The diagnosis should be given if at least five of these criteria are fulfilled. These criteria are to be separated expressly from the use of professional and other private Internet use. Rehbein et al. (2013) point out that thus far these criteria have not been empirically verified, but express hope that this specification of criteria could be a starting point for further research into the development of appropriate diagnostic instruments.

It is thus clear that many different perspectives are accommodated within the research on (problematic) Internet use. Indeed, general sociological factors (Livingstone 2009) seem to play just as much a role as generation processes concerning the different media socialization of "digital natives" and "digital immigrants" (Shell Deutschland Holding 2011). There is also some debate regarding the correct diagnostic classification of the behaviours related to PIU, i.e., either as an addiction or other Axis I disorder, as in the case of pathological gambling and other behavioural addictions (Mann et al. 2013). As important as this debate may be, it contributes rather little to improving understanding of the very different profiles of use and application that the Internet offers. The position of the therapist, a possible "digital immigrant" himself, offers a unique challenge. The task consists, after all, in providing adequate therapeutic access for the clients without lapsing into "moral panic" or playing things down.

Table 12.1 DSM-V criteria of the Internet gaming disorder

1. Preoccupation with internet games
2. Withdrawal symptoms when Internet gaming is taken away (irritability, anxiety, sadness)
3. Tolerance—the need to spend an increasing amount of time engaged in Internet games
4. Unsuccessful attempts to control participation in Internet games
5. Loss of interest in previous hobbies and entertainment
6. Continued excessive use of Internet games despite knowledge of psychosocial problems
7. Has deceived family members, therapists, or others regarding the amount of Internet gaming
8. Use of Internet games to escape or relieve a negative mood
9. Has jeopardized or lost a significant relationship, job, or educational or career opportunity because of participation in Internet games

In addition to the DSM-V criteria, which possibly will facilitate more comparable research findings, the cognitive behavioural model (Davis 2001) also seems to be helpful in classifying the disparate phenomena related to problematic Internet use.

12.1.2 The Cognitive Behavioural Model of Davis: Specific and Generalized Problematic Use of the Internet

Davis (2001) suggested the need to make a distinction between a ***specific*** and ***generalized*** form of Internet abuse within his cognitive behavioural model of PIU. For people with a generalized Internet abuse problem, the medium of the Internet is proposed as a stressor that ultimately leads to the exacerbation of particular behaviours. On the other hand, people with a specific Internet abuse problem, use the Internet merely as another medium through which the addicted behaviour is shown. As a consequence this kind of addiction would also be observed in the absence of the Internet (e.g. addiction to online and offline pornography). Another example of this constitutes a person suffering from a pathological gambling addiction, who gambles both on slot machines as well as online. This model is illustrated in Fig. 12.1. The criteria identified in this model of problematic Internet use closely resemble those for "Internet Gaming Disorder" discussed above, but with a stronger focus on cognitive symptoms. Of particular importance is the belief that the Internet is the only place where one feels good.

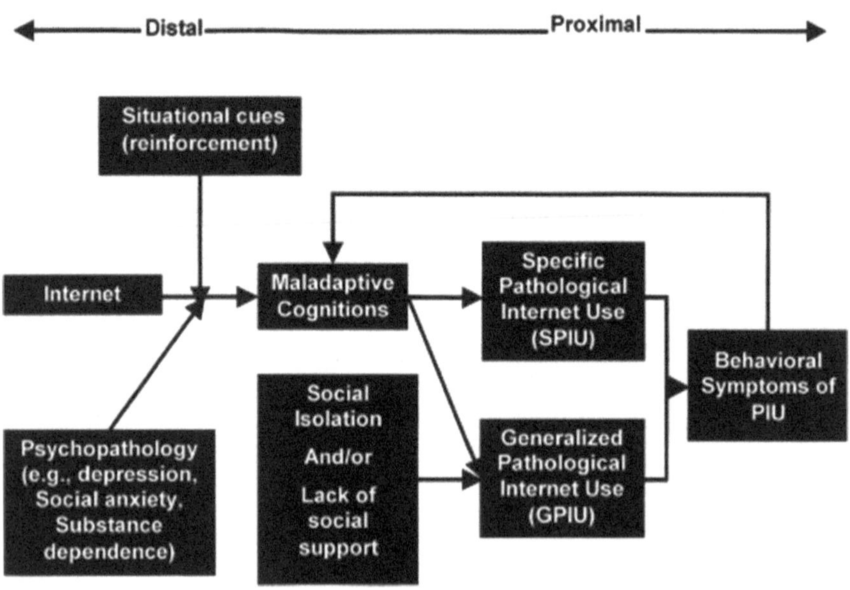

Fig. 12.1 Cognitive-behavioral model of problematic internet use (Davis 2001, p. 190)

Another advantage of the model is that the author tries to incorporate healthy use of the Internet. A continuum between a healthy and a pathological use of the Internet is suggested. The distinction between functional and dysfunctional Internet use can only be made on a case-by-case basis, taking the model into account (see Fig. 12.1; Davis 2001). This model has received empirical support (Caplan 2010; Montag et al., 2014), however, the role of Internet-related cognitions still requires clarification (Li et al. 2010).

The cognitive behavioural model is very helpful since it provides a theoretical framework for research as well as therapy (Davis 2001). Further research is needed to determine the extent to which the specific and generalized forms of pathological Internet use necessitate different therapeutic approaches.

Figures on Internet usage for the general population and the prevalence of problematic Internet use are presented below. Following this, treatment alternatives are introduced. In the context of a case study, the short intervention, "Compass", will be detailed, before a final discussion of the empirical research on the efficacy of existing approaches to the treatment of problematic Internet use.

12.1.3 General Use of the Internet

12.1.3.1 Adolescents and Young Adults

According to surveys by the Statistisches Bundesamt (official statistics office; 2013), 77 % of households in Germany had Internet access in (2011), of which 98 % had a broadband Internet connection. Among adolescents, 87 % said they had Internet access in their own room (MPFS 2013). The everyday importance of the Internet is reflected in the fact that 68 % of 12 to 19-year-olds reported that they surfed the Internet daily, and 91 % at least several times per week. The sociodemographic variables, gender and educational background, were not significant. In contrast, the frequency of Internet usage increases with age: among 12–13-year-olds only 48 % use the Internet daily, while this true of 69 % of 14–15-year-olds. Adolescents estimated their total daily usage to be 131 min per day. While no gender differences were observed, educational differences did emerge: adolescents who attended Elementary school, spent more time on average on the Internet than pupils of Grammar Schools (157 min vs. 124 min) with Technical school pupils being in the middle of the observed usage (134 min).[1]

The use of the Internet on mobile phones or Smartphones increased 20 % from 2011, up to 50 % in 2012, with no observed gender differences. Figure 12.2 provides information on the types of use and related gender differences.

[1] In Germany four kinds of secondary schooling exist: (1) Grammar school (Gymnasium) until grade 12, qualifying for university. (2) Technical school (Realschule) until grade ten. (3) The least academic Elementary school (Hauptschule) until grade nine or ten, and (4) the Comprehensive school (Gesamtschule), which is not included in the statistic.

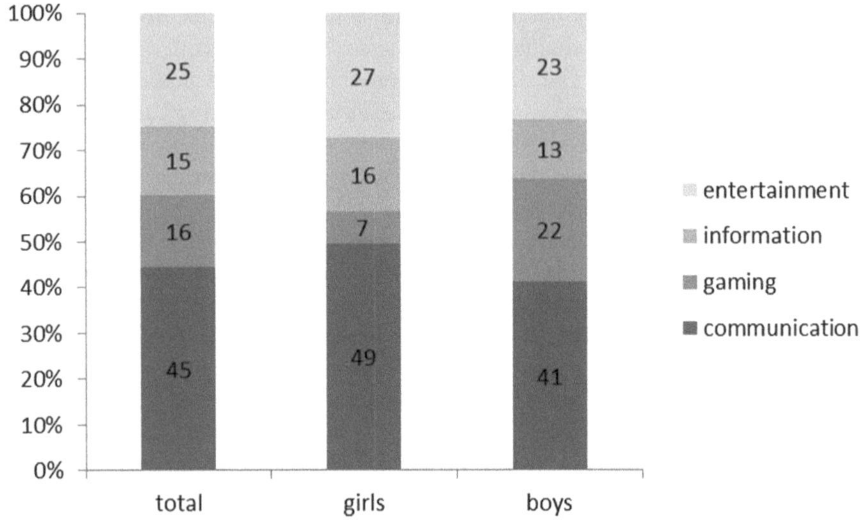

Fig. 12.2 Frequency of Internet usage depending on gender (MFPS 2013)

Girls use the Internet more for electronic communication and less for gaming compared with boys. Similar results are observed in other representative surveys on Internet use (Czaika 2011; Shell Deutschland Holding 2011). Evaluations using factor- and cluster- analyses (Leven et al. 2011) classified adolescent users into so-called "Gamers" (25 %), "Function Users" (17 %), "Multi-users" (34 %) and "Digital Networkers" (24 %). In addition to game-play, the "Gamer" group are also unique in having a markedly lower use of digital communication forms (e-mail, social networks). "Digital Networkers" showed the opposite pattern: they used all forms of digital communication, especially social networks, and spent the most time on the Internet (14.6 h per week). This group also often surfs the Internet at random and more than half (53 %) of them never play online games. The "Function Users" use the Internet very specifically, e.g. for shopping and information searches, while the "Multi-users" are varied and non-specific in their Internet usage; however, it is worth noting that this group also seldom plays online games (Leven et al. 2011). Besides this, the groups also differ in terms of other features. The "Gamers" are younger, mostly male, often have lower socio-economic backgrounds and are more likely to live in conurbations. The "Multi-users" are also mostly male, often come from higher socio-economic backgrounds and often live in cities. Women form the majority of the "Function Users" and the "Digital Networkers", the latter coming primarily from more rural areas. After the "Gamers", the (majority female) "Digital Networkers", represent the second-youngest group. Given these clear age delimitations, Leven et al. (2011) propose a model for the development of Internet usage from online gaming, through "network use", up to a purely functional use. However, the authors concede that further research is needed on this. However, the growing popularity and importance of the Internet among adolescents and young adults cannot be denied.

12.1.3.2 Usage in the General Population and Generation Differences

Both in terms of media usage and expertise, significant generational differences have been observed, with younger users being more advanced and possessing a "pioneering role" (Hurrelmann and Quenzel 2012, p. 197). Furthermore, Hurrelmann and Quenzel (2012) show that adolescents have a more confident and more natural handling of the Internet and smart phones. While differences in usage traced to socio-economic background have nearly levelled out (Christakis 2010; MFPS 2013; Ridder and Engel 2010), this is not true for age differences. Internet usage is on the increase among older people in Germany. While the proportion of over 60 s using the Internet, the so-called "silver surfers", was 28.2 % in 2010, this proportion has increased to 42.9 % in 2013 (van Eimeren and Frees 2013). Of note, the average duration of daily Internet use decreases with increasing age (see Table 12.2).

All user groups show longer duration of Internet use from Monday to Friday (185 min) than on the weekend (130 min). Older users are less varied and explorative in their usage of the Internet than their younger contemporaries. Users over the age of 50, typically restrict themselves to e-mail communication and targeted information searches (van Eimeren and Frees 2013).

In contrast, usage of the Internet for communication (e-mail, social networking) ***and*** for entertainment is more common among younger groups. In view of how common Internet usage has become among the general population, one can ask whether a division between the 'real' and virtual/digital world still makes sense. At least for the younger age groups, such a separation seems increasingly artificial. The importance of both quantitative and qualitative differences in Internet use is illustrated by generational differences in how personal data is handled online. Younger people show a much greater willingness to reveal personal data, for instance on social networks, than older people (Leven et al. 2011). This may of course be interpreted as an expression of greater carelessness in the handling of personal information, however it may also be reflective of a fundamentally different relationship towards new media (Leven et al. 2011). Although a substantially higher number of younger users report having been "bullied" on the Internet, fears and concerns about negative consequences of Internet use, like data abuse, are more common among older people (Czaika 2011). Further research will show whether these trends will be confirmed and/or reinforced (Fig. 12.3).

Table 12.2 Average of minutes per day spent online (according to van Eimeren and Frees 2013)

Age group (years)	Minutes per day online-time (average)
14–29	237
30–49	168
Over 50	116
Total	169

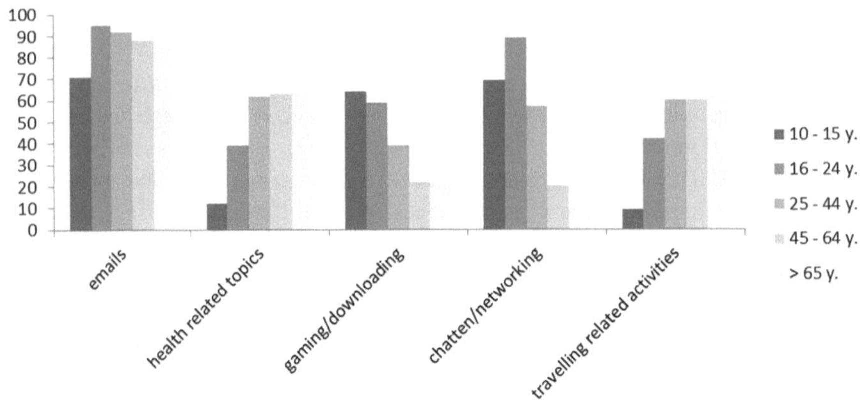

Fig. 12.3 Internet usage depending on age groups (Czaika 2011)

Table 12.3 Age structure of internet users using the internet several times per week in Austria and Switzerland (Bundesamt für Statistik Schweiz 2010; INTEGRAL 2012)

Age (years)	Percentage of regular internet users	
	Austria	Switzerland
14–19	97	94.9
20–29	95	94.5
30–39	94	90.6
40–49	89	83.9
50–59	83	77.7
60–69	66	62.2
>70	34	32.2

12.1.3.3 Use of the Internet in Austria and in Switzerland

In Austria, 79 % of households had Internet access in 2012 and 85 % reported the possibility, in principle, of having a connection with the Internet (e.g. through workplace, school, university etc.) and repeated use of the Internet was nearly equally frequent for men and women: 54 % versus 46 %. Specifications on the age structure of the use are reproduced in Table 12.3 (Integral Markt- und Meinungsforschungsges. m. b. H. (INTEGRAL) 2012).

In Switzerland 85 % of households had Internet access in 2012, but otherwise hardly any differences in usage behaviour exist between the countries (Bundesamt für Statistik and Schweiz 2010; Czaika 2011; INTEGRAL 2012).

12.1.3.4 Summary

The Internet has become a part of everyday life in Germany and this is also true for Austria and Switzerland. Depending on age, the usage and subjective importance of the Internet apparently seem to differ. While rather specific applications and functions of the Internet seem most interesting for older people, for younger people the Internet has become an important component of experience and behaviour. Young people more commonly share private information on the Internet and demonstrate a rather carefree attitude toward the Internet as a medium. Just as everyone in school knows that a student is single or a member of a sports team, this is similarly communicated to the web community, with little concern for data protection. In addition, the proportion of adolescents with personal access to the Internet is on the rise and, correspondingly, uncontrolled Internet access is increasing. In this regard, the existing evidence suggests that absent parental control increases the risk for problematic Internet use (Lin et al. 2009).

12.2 Prevalence of Problematic Internet Use

In Germany, two representative population studies of the prevalence of problematic Internet use are of particular relevance. In a representative study by the University Hospital of Mainz, the prevalence of Internet-dependent people was found to be 2.1 % (Müller et al. 2013).

However, the study carried out by order of the federal health ministry, "Prevalence of Internet dependence (PINTA)" (Rumpf et al. 2011), identified 1 % of the whole population as Internet-dependent. The results of the PINTA study fit well with the findings from international studies. For instance, prevalence rates between 0.3–0.7 % arose for the USA (Aboujaoude et al. 2006) and of 1 % for Norway (Bakken et al. 2009). For Switzerland, the prevalence is estimated at 2.7 % (Eidenbenz 2011) and for Austria, 2.4 % (the estimates for Austria vary between 1.95 and 2.81 %; Batthyány and Pritz 2009).

In summary, the prevalence for computer game dependence in Germany is taken to be between 0.9–1.7 % (Rehbein et al. 2013). For cybersex addiction, defined as the excessive use of sexual material, the prevalence estimates vary as a function of the investigation methods used, between 1–8.6 % (Eichenberg and Blokus 2010).

Furthermore, the PINTA study shows that the prevalence of Internet dependence is negatively correlated with age; the prevalence is higher among younger users. Furthermore, females are more often affected and significantly more often use social networks (Rumpf et al. 2011). If you compare these results to the findings from Leven et al. (2011), a more fine-grained picture emerges. Here, the problem of an excessive or dependent use of the Internet seems to concern mostly the groups of "Gamers" and "Social Networkers". With computer game dependence and cybersex addiction, problem users are most often male, while female users more often have difficulties controlling their use of social networks (Eichenberg and Blokus 2010; Rehbein et al. 2013; Rumpf et al. 2011).

Rehbein et al. (2013) point to the high degree of comorbidity observed when investigating problematic Internet use. In particular depressive disorders, anxiety disorders and attention–deficit disorder appear comorbid with problematic Internet use. This high degree of comorbidity is one reason why the stand-alone diagnosis "Internet addiction" is viewed critically by some researchers (e.g., Kratzer and Hegerl 2008). According to Rehbein et al. (2013), the spectrum of comorbid disorders present in Internet addiction, bear resemblance to those comorbid in people suffering from substance-related addictions and also pathological gambling.

Longitudinal studies on the course of Internet Addiction or problematic Internet use are still rare. For dependent online video game players, Van Rooij et al. (2011) found that after one year 50 % of the random sample still fulfilled the criteria for addictive gaming, while Gentile et al. (2011) showed symptoms can continue to exist for years later. For this, further research is urgently needed.

In sum, it can be ascertained that the prevailing majority of Internet users show no pathological patterns of use. However, it seems that female users of social networks have an increased risk of developing an problematic Internet use. Furthermore, it can be stated that both healthy Internet users and those with problematic Internet use represent rather heterogeneous population groups with different issues and problems. This needs to be kept in mind when offering treatment and consultation.

It should also be taken into account that, particularly for younger people, the Internet forms a natural and integral component of their social environment. Differences between the generations and lessons gleaned from the debate on television should be considered in the design of treatment and counselling services for Internet Addiction. On one hand this is needed to counteract preventable reactance from the target group, which presumably will consist primarily of younger people. On the other hand, it is necessary, as stereotypes about young people and their supposed usage of the Internet can lead to other more important problems being overlooked, e.g. depression and/or social anxiety. The distinction between "real" and "virtual" in this case seems of little help. Empirical findings suggest that an essential function of social network usage is the intensification and maintenance of off-line friendships/relationships (Ellison et al. 2007; Kraut et al. 2002). However, socially nervous adolescent users, in-keeping with the model of Davis (2001), seem to have other expectations of online communication; socially anxious adolescents value the importance of online communication on intimate topics significantly higher than their non-anxious peers (Valkenburg and Peter 2007).

12.3 Treatment Programmes in Germany

In Germany it is predominantly the addiction help system that provides and further develops prevention, counselling and treatment for Internet Addiction (Petersen and Thomasius 2010). On the relevant websites in the field of addiction support, in addition to information, there are self-tests to assess one's own Internet use, consultation chats, e-mail contact details, as well as brief interventions or tools to facilitate autonomous change. University institutions are also engaged in research on problematic Internet use (te Wildt 2011). In addition to this professional orientation, homepages of people and their families who are affected by Internet Addiction promote the issue and its accompanying problems, provide information and recommend contact with therapists, e.g. www.rollenspielsucht.de or also www.onlinesucht.de. A professional association has also been developed ("Media addiction association"/"Fachverband

Medienabhängigkeit e.V.") and the German Association for Psychiatry, Psychotherapy and Psychosomatics [Deutsche Gesellschaft für Psychiatrie, Psychosomatik und Nervenheilkunde (DGPPN)] has founded a group for the investigation and nosological classification of Internet addiction.

Besides offers of consultation from outpatient addiction specialist agencies, which also do preventive work on media competence in schools, the outpatient clinics of psychiatric hospitals, psychosomatic hospitals and rehabilitation clinics (addiction and psychosomatics) predominantly offer treatment (te Wildt 2011).

Research of online manuals and treatment concepts in Germany suggests that there currently exist cognitive-behavioural therapy-oriented concepts, attachment theory- and depth psychology based- approaches. The latter also integrate behavioural therapy-oriented technologies and interventions (Petersen and Thomasius 2010).

The website of the trade association for media addiction lists numerous support services in Germany, Austria, Switzerland and Luxembourg: http://www.fv-medienabhaengigkeit.de. It is beyond the scope of the present chapter to review these services in detail. Based on their pioneering nature and/or reputation, three services have been selected for discussion below.

12.4 Lost in Space/Café Beispiellos Berlin

Since 1987, the Café Beispiellos has existed in Berlin as a consultation service for addicted gamblers and their relatives, with sponsorship from The German Caritas Association Berlin (Caritasverband des Erzbistums Berlin e.V.) As of October 2006, an additional service specifically for Internet and computer addicts has been provided, called "Lost in Space". The objective of "Lost in Space" is to compile alternatives to computer use and/or to develop better time management thereof. The service offers various activities to facilitate these aims. Furthermore the consultation service offers counselling for individuals, couples and families. This service can be accessed via: http://www.computersucht-berlin.de.

12.5 Schwerin Media Addiction Counselling

The counselling centre for excessive media use and media addiction emerged in November 2006 as a cooperation project between the Mecklenburg-Vorpommern Evangelical Addiction Help and the Schwerin Helios medical centres. The offer is directed at all age groups and can be used by telephone, in writing, or in person. The objectives are to find ways out of addiction and to understand the underlying causes. The teaching of media literacy to relatives is seen as an integral part of the offer. The website is available at http://suchthilfe-mv.de/vermittlung/vermittlung3/allgemeines.php.

12.6 Computer Game Addiction Outpatient Clinic of the University Hospital in Mainz

"Sabine M. Grüsser-Sinopoli Ambulanz für Spielsucht" of the University Hospital of Mainz was opened in 2008. Initially created as a 12-month pilot project, computer game- and Internet-addicted people are treated on an outpatient basis, as well as pathological gamblers. The offer includes free telephone consultation for concerned friends and relatives, as well as elements of behavioural therapy talks. In addition to individual meetings, group therapy takes place over 20 sessions (Wölfling 2009). There also exists a cognitive-behavioural treatment manual for computer game- and Internet- addiction (Wölfling et al. 2013).

In a little more detail, we will now look at the treatment and counselling services of the LVR Clinic of Bonn. The short intervention, "Compass", used at the LVR clinic will be described inter alia.

12.7 Treatment and Consultation Services for Media Dependence at the LVR Clinic in Bonn

In 2004 the LVR Clinic in Bonn, in cooperation with the specialist agency for addiction prevention, *update,* started the special treatment service "Bonn model—young addiction", with financial support from the city of Bonn, in view of rapidly increasing prevalence and treatment demands (Dau et al. 2008). First, this outpatient–inpatient treatment service was directed at young cannabis and party drug users over the age of 18. The cognitive-behavioural therapy approach includes content based on a developmental-psychology model for the development of addiction. Moreover, learning to adequately handle alcohol and drugs can be understood within the context of age-specific developmental tasks (Havighurst Havighurst 1972). A failure to accomplish these developmental tasks bears an increased risk of later dependence (Hurrelmann and Quenzel 2012). Therefore, this treatment approach orientates itself more in-line with psychosocial development, than substance consumption. Based on this approach, an independent consulting and treatment service for media dependency was established in 2009 in the outpatient clinic of the Department of Addictions and Psychotherapy, at the LVR Clinic in Bonn. This treatment service also encompasses the diagnosis and treatment of comorbid disorders. The service is directed at all age groups. It must be noted that the development of identity and autonomy are central issues in adolescence, and that the Internet, with its information and communication possibilities, offers various opportunities for the fulfillment of these needs (Lei and Wu 2007). For some individuals, the risk of failure or non- age appropriate coping with developmental tasks may lead to an increased risk of developing a problematic Internet usage, e.g. if autonomy and/or demarcation demands are acted out too excessively on the Internet.

12.8 Is Abstinence a Sensible Therapy Objective?

A complete abstinence from the Internet may only prove a realistic—or even useful- therapeutic aim in very exceptional cases. The Internet is too pervasive in important areas of life for this aim to be broadly feasible. The treatment approach of the department outpatient clinic is thus characterized by one principal goal: openness. Depending on the patient's planning and goal setting, the entire renunciation of a specific application or a reduction in the use thereof may be agreed upon as potential objectives. The patient's motivation for this goal setting needs to be taken seriously. The importance given to Internet activities in the life of the patient, both currently and in the post-treatment future, requires discussion. The definition of very concrete contingents of Internet use and early warning signs for relapse into older, problematic online behaviour patterns, also appear to be essential.

The role of self-esteem problems, stressful life events and family conflicts in relation to problematic Internet use has been highlighted by many studies (e.g. Rehbein et al. 2010). This also needs to be considered in diagnosis, case-planning and treatment. If necessary, inclusion of relatives or family members in the treatment process may be useful. Clinicians should be mindful, however, that the patient's possible need for increased autonomy may result in a potential conflict endangering positive treatment outcomes.

In addition to outpatient services, there also exists the possibility of acute inpatient treatment for those strongly affected by Internet Addiction and, where applicable, comorbid mental disorders, e.g. heavy depressive disorders. There is also the possibility of establishing email contact via www.mediensucht-bonn.lvr.de.

There were 75 accepted requests for access to this service at the time of going to press. In about 35 cases, continuous treatment was recorded, with more than five single-therapeutic sessions. In three cases, inpatient treatment was necessary due to the presence of an acute depressive disorder. In this context, Teske et al. (2013) note that many Internet abuse services record significantly fewer help-seekers than would be expected based on prevalence estimates. In the opinion of the authors, this may be explained by the low awareness of the disorder and those problems connected with excessive use of the Internet. The extent to which demand for treatment is consistent with the prevalence identified so far, however, remains to be seen. In the following sections, the treatment approach as implemented in the LVR Clinic in Bonn, as well as the theoretical considerations thereof, are outlined in more detail.

12.9 Internet-Related Procrastination Meets Developmental Tasks

Helpful to the analysis of Internet Addiction is the consideration of motivational-psychological approaches (e.g. the research field of procrastination (Steel 2007), in which postponing behaviour is explored). Such approaches can facilitate a better understanding and treatment of problematic Internet use. The relationship between procrastination and problematic Internet use is empirically based (e.g. Lim 2002; Thatcher et al. 2008). Lavoie and Pychyl (2001) found a strong relationship between trait procrastination, negative emotions and Internet procrastination: 51 % of their study participants stated that they frequently surfed the Internet at work, instead of beginning the execution of tasks, and spent 47 % of their whole time online procrastinating, although trait procrastination was not increased among this sample. For the usage of social networks and online gaming via the Internet, an association between usage time, procrastination and life satisfaction can be shown: heavy users showed increased procrastination and lowered life satisfaction (Hinsch and Sheldon 2013). Another study showed that more than 3 h at work each week are spent on the Internet performing non-work related assignments (Zanna 1996). Such procrastination behaviour via the Internet is associated with negative feelings, stress, lapses in productivity, and negative self-evaluation (Davis et al. 2002). It seems that some individuals use the Internet as an avoidance activity in order to be distracted from a stressful event, task or stream of thought. In an online study of 1,399 participants, a strong connection between procrastination, problematic Internet usage and the flow experience was found (Thatcher et al. 2008). The flow experience (Csikszentmihalyi 1997) describes an intrinsically motivated state occurring when task difficulty and the self-perceived abilities of a person to successfully work on this task are almost identical. The flow experience is also characterized by a clear objective control over the activity, immediate feedback, concentration and focusing, loss of conscious self-experience, loss of sense of time and is experienced predominantly as pleasant (Csikszentmihalyi 1997). Many activities that are carried out with—and on—the Internet have the potential to convey a flow experience (Thatcher et al. 2008). Accordingly it can be expected that an individual who avoids a stressful task by using Internet activities, e.g. online gaming, could get caught up in the pleasant 'flow' of this avoidance activity. As a consequence the gaming session, begun as an avoidance behaviour, will probably last much longer than originally planned. Indeed, this vicious circle is supported by research findings (Thatcher et al. 2008; Wan and Chiou 2006).

To further illustrate this, we outline the following case study: A 25-year-old man with a depressive disorder and problematic Internet use sought therapy. He had failed to successfully apply for a job upon completion of his education. Moreover, he failed to sustain a long-term relationship. The patient's mother accompanied him to seek help. He stressed the importance of being head of an online role-player guild. This game-related success, in which the patient had invested a lot of time, money and commitment, were ultimately more important to him than the game experience itself. On the other hand, he failed to deal with his every-day mail, and had failed to report to the unemployment office, with the consequence that he had lost his home shortly before coming into our clinic. Due to the belief that he could only be truly successful in his online game (Davis 2001), the patient had used Internet games to avoid confronting important job-related tasks, which he experienced as aversive (Steel 2007). Instead he satisfied his need for achievement through his role as head of the online guild. The actual playing of the online game was rather of secondary importance, here.

This Internet-based procrastination fits within Davis' (2001) cognitive-behavioural model for problematic Internet use, and gains therapeutic relevance through an improved understanding of the mechanisms underlying problematic Internet use. In the following section we will address how problematic Internet behaviour mirrors the attempt of the patients afflicted to treat their problems by using the Internet.

In addition to the short therapeutic intervention "Compass", outlined in more detail below, interventions for the treatment of procrastination, as developed e.g. by Höcker et al. (2013) may also be useful. This notion is supported by the findings of Montag et al. (2010). They reported that low self-directedness, as defined by the Temperament and Character Inventory (Cloninger et al. 1993) is a better predictor than neuroticism for problematic Internet use. Self-directedness accompanies the feeling of having control over one's own life, an active coping strategy in solving everyday problems, and self-acceptance of one's own personality. Procrastinators typically lack these personality characteristics. Of note the association between problematic Internet use and low self-directedness has also been replicated in a large group of "*Counterstrike*" Gamers (Montag et al. 2010) and also cross-culturally in seven countries, including the continents Europe, Asia and South America (Sariyska et al. 2014).

12.10 Goals Against "Clicks": The "Compass" Short Intervention for the Treatment of Problematic Internet Use

> The tragedy of life doesn't lie in not reaching your goal. The tragedy lies in having no goals to reach.
> (R. H. Smith)

The following sections outline the development of the "Compass" program, which has originated from the results and experiences of the "Bonn model—young addiction" treatment program. We also outline how this program relates to a cognitive-behavioural approach to the treatment of problematic Internet use, as described by Davis (2001).

12.11 Development of the "Compass"

In the context of research dealing with cannabis patients (Aden et al. 2011), it was observed that patients sought treatment more because of massive external pressure, rather than an intrinsic motivation. This raised the question of how patients could be better supported to comply with the therapy. The basic idea was to integrate the positive effects of therapeutic homework (Kazantzis et al. 2005) as effectively as possible in the inpatient treatment and to associate it with experiences of short interventions (Haug et al. 2010).

12.12 Background: Self-regulatory Theoretical Underpinning

In the development process of "Compass" several theories including self-management therapy (Kanfer et al. 2000), expected self-efficacy (Bandura 1977), reactance research (Brehm 1966), Motivational Interviewing (Miller and Rollnick 2002) and solution-focused therapy (DeShazer 1989) were considered.

Box 1: The four basic principles of the "Compass"

(a) Motivation is assumed in the patient; its catalyst is the task of the therapist,
(b) the reduction of demoralization through the mediation of experiences of success is crucial for the development of expected self efficacy,
(c) a reduction of reactance is achieved through acceptance, transparency and freedom of choice, and
(d) the mediation of problem-solving knowledge to improve the self-management skills.

One of the basic assumptions in self-management therapy is to support people in creating an *active* lifestyle. In order to maximize *personal freedom,* decision-making authority and freedom of action should be attained in place of habits and stereotyped behaviour patterns (Kanfer et al. 2000). Instead of using the Internet for procrastination behaviour, e.g. in avoidance of more stressful challenges, self-management therapy aims to enable people with alternative behaviour strategies. If the therapy is successful, the individual can then choose between procrastination and task-solving behaviour, which provides an increased level of personal freedom due to more accessible behaviour strategies. Self-regulation and self-control represent the central approaches of self-management therapy for the "Compass" program. Self-management refers to the overall ability of a person to control and/or change their own behaviour under explicit or implicit use of specific strategies (Kanfer et al. 2000). The processes involved in this control are described in the concept of self-regulation, which comprises the steps of "introspection", "self-assessment" and "self-reinforcement". Insofar as problematic Internet use is defined as "an individual's inability to control his or her use of the Internet, causing marked distress and/or functional impairment" (Shapira et al. 2000, p. 267), a key focus in therapy is the strengthening of self-control. In the self-management approach, and therefore also in the "Compass" intervention, self-control is understood as a special case of self-regulation. A person displays self-control if, when experiencing a conflict, that person exhibits a behaviour that would not be expected from the perspective of short-term contingencies (Kanfer et al. 2000). These conflicts typically arise when short-term contingencies suggest avoidant behaviour, e.g. not learning for an exam (short-term contingencies: relief, spare time), but the long-term contingencies (failing the exam, receiving poor marks, not getting a degree) demand task-solving behaviour. In such a case, the individual is forced to decrease negative emotions, like fear, anger, or boredom, and to increase or activate problem solving behaviour. Two basic types of self-control can be distinguished and are depicted in Box 2.

Box 2: Two basic types of self-control (Kanfer et al. 2000)

1. **Resisting a temptation**:
   An enjoyable activity, like playing online games on the Internet is forfeit in order to achieve other goals, like getting a degree or fulfilling obligations.

2. **Heroic behaviour (enduring an aversive situation)**:
   Short-term negative consequences will be tolerated to cause long-term positive results, e.g. going to the dentist, learning for an exam, enduring craving to regain control over the Internet usage etc.

Here, self-control is understood as actions of people in conflict situations, without referring to individual differences or personality traits. In addiction, self-control is distinguished into decisional or protracted self-control. The former can be described as a conflict which is ended through a decision by the individual themselves (decisional self-control), the latter describes an aversive condition that has to be endured, or a temptation resisted, for a longer period (protracted self-control) (Kanfer and Gealick 1986). As a rule, protracted self-control is particularly difficult for most people, but is especially important in the treatment of problematic Internet use, as total abstinence from the Internet is seldom a realistic goal. The way in which "Compass" supports clients' self-regulation is outlined below. From a theoretical point of view, we will next consider the importance of self-relevant objectives in achieving self-regulation (Fig. 12.4).

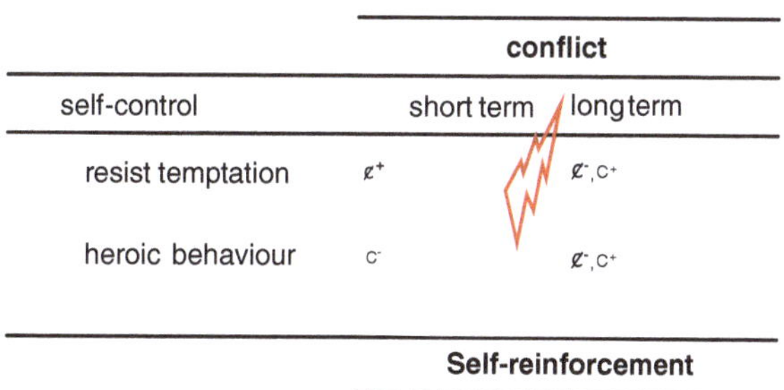

Fig. 12.4 Self-control as a special case of self-regulation with notation of the SORCK-model (Kanfer et al. 2000)

12.13 Background: Underpinnings from Goal Setting Theory and Motivation Psychology

In goal-setting theory, objectives are defined as an object or objective of an action and can be identified by content and intensity (Latham and Locke 2013; Locke and Latham 2002). Here, the content of an objective refers to the target object or the state to be achieved, e.g. get up on time. The objective intensity consists of the effort, the position of the target in the personal hierarchy of objectives and the degree of commitment of the individual to achieve their objectives (Latham and Locke 2013). Specific objectives are typically more successfully reached compared to unspecific or summary objectives. In terms of percentage, the performance improves through specific formulation in ranges from 8 up to 16 % (Wood et al. 1987). So the formulation of a specific objective such as "get up at 06:30 h" is more favourable than the more vague "get up on time". Furthermore, important variables include freedom of choice, task difficulty, effort, perseverance, and solution strategies (Locke and Latham 2013). Simplified and schematically, the relationship is shown in Fig. 12.5.

These findings from goal-setting theory can be easily reconciled with the self-management approach. Also, according to Kanfer et al. (2000), therapy objectives have to be (a) specific, (b) realistic, (c) attainment must, in principle, be subject to client control and (d) include self-reward in case of reaching the goal. These principles of goal formulation are incorporated in the "Compass" program. In general terms, objectives take on a directive, attention-controlling function for behaviour. Under the aforementioned conditions, objectives increase motivation, as well as perseverance, and they encourage the acquisition of task-relevant knowledge (Locke and Latham 2002). The benefit for psychotherapy, therefore, is clear, especially if the determining parts of the definition of "flow experience" are considered (Csikszentmihalyi 1997). Hence, during goal formulation in the "Compass" program, it is important to use cognitive methods to teach patients both to gain a realistic perception of their own abilities, as well as to set goals at a reasonable level of difficulty.

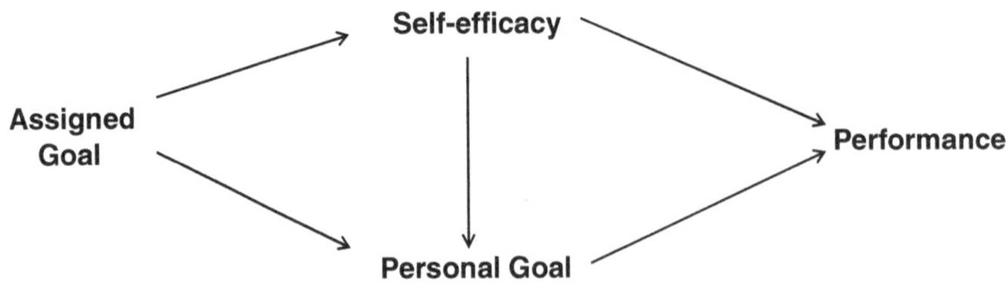

Fig. 12.5 Moderators and mediators between goal setting and performance according to Locke and Latham (2002)

12.14 Background: Motivational Underpinnings in a Nutshell

At first it seems difficult to understand that people with problematic Internet use are willing to accept negative consequences in order to spend more time on the Internet. Over the course of this development, it may be possible to observe that Internet use becomes increasingly important for the person, while other activities/objectives become steadily more repressed. From classical operant conditioning research it is known that immediate positive reinforcements have a profound behaviour-controlling effect. When an individual spends a lot of time on the Internet and takes a lot of joy and/or relaxation from it, s/he will increasingly carry out this behaviour. Moreover, the individual will probably feel very good due to the Internet and may connect the unspecific objective "to feel good" with Internet use. Internet use then becomes instrumental to the objective "to feel good", and so a strong association will originate between "Internet" and "good mood" (Fishbach et al. 2004). Thus, the motivational value of the objective "to feel good" is transferred onto the means for goal achievement, in this case Internet use. This process, called "means valuation", is well supported empirically (Brendl et al. 2003; Ferguson and Bargh 2004; Fishbach et al. 2004). In the long term, this effect can have the result that the original means for goal achievement (Internet use) becomes an objective in itself (Köpetz et al. 2013).

Furthermore, it is known from research that goal pursuit is dependent on resources and that intra-psychological resources for self-regulation can at times be exhausted (Baumeister et al. 1998; Hagger et al. 2010). This can be illustrated as a muscle being trained and strengthened through use, but which can also be temporarily exhausted. It can be assumed that such resource exhaustion occurs over time, especially when a person is practicing self-monitoring techniques, in particular with "protracted self-control". Clearly, relapses and setbacks can be explained by such a lack of resources. People react to resource limitation by using their resources strategically, depending on the individual significance and importance of objectives (Kruglanski et al. 2012). If several objectives are active at the same time, e.g. desire for Internet use and learning for an exam, people protect the currently most important goals through highly automated inhibition of competing goals (Köpetz et al. 2011). The most important current goal is followed, in spite of competing goals, and hence also in spite of possible negative consequences. Such tendencies may generally play an important role in behaviour related to addiction (Köpetz et al. 2011, 2013). Over time an original means of goal achievement can become in itself a goal that, if the positive emotion is strong enough, has the potential to increasingly dominate other goals. Under certain conditions afflicted persons cannot get out of this vicious cycle without help, even if they want to do so.

If we follow the assumptions made above, two main approaches should be pursued in therapy: on one hand, self-regulation resources should be maximised to prevent exhaustion. On the other hand, the person should have as many specific, moderately difficult and problem-behaviour related competitive objectives as possible, and the means to achieve them. It is described how such objectives can be established and how self-regulation resources can be protected.

12.15 Cure for Problematic Internet Use: How the "Compass" Helps Beating Clicking

First, implementation of the "Compass" is described in its original form, as conceived for an inpatient setting. Second, other possible applications are outlined for an outpatient service. Of note, in inpatient therapy the patients receive an additional daily therapeutic contact of maximum 5 min duration. During the "Compass" session, a short behaviour-related assignment (one goal) is formulated daily by the patient himself. In the "Compass" discussion the next day, monitoring of the performance and achievement of the assignment is carried out, based on a scale rating (scale 0–100), as well as the establishment of a new goal. Through this, the "TOTE" scheme (Test-Operate-Test-Exit) should be taught as a simple problem-solving algorithm (Miller et al. 1960).[2] As the patient is responsible for setting the objective, the therapist provides support only in shaping and operationalising the specific goal. In doing so, the therapist works to promote motivation and to reduce reactance as far as possible. As a general instruction, the patient is told that the goal must be achievable within a day or a weekend and the implementation must also be measurable. The goal is written down, read to the patient again and the patient is positively reinforced, i.e. praised, for the formulation. After that the therapist asks whether the patient requires assistance to achieve the goal and briefly intervenes if necessary.

12.16 Goal Openness and Instruction for the Patient

Goal openness on the part of both the therapist *and* the patient should provide the greatest possible acceptance of the client's potential, should increase compliance with—and therefore decrease reactance to—therapy and behavioural change. The patient is given the suggestion to formulate the goal very positively and specifically (positive rule; e.g. "get up at 6.30 am" instead of "do not oversleep"). In giving this instruction, the therapist also orientates himself toward the resource-oriented approach (Haug et al. 2010). If no goal can be formulated, it is suggested that the goal should be to find a goal for the next session. A time limit is imposed in such a situation to avoid excessive problem orientation. This saves both the resources of therapist and patient and provides a feeling of "Less can be more".

[2] The TOTE scheme is especially useful in the early stages of therapy due to its simplicity, although it does not do full justice to the complexity of the underlying process.

12.17 Intervention During the "Compass"

In a way, "Compass" may be considered a special form of the well-established therapeutic 'homework' as known from behavioural therapy. Issues that arise in individual or group sessions can be integrated into the "Compass" and vice versa. "Compass" intervenes if the patient sets himself inaccessible or unspecific goals, or if he proceeds with attainment scaling too strictly. In such a case, the therapist will use cognitive techniques (socratic dialogue) to dispute the strict judgement of the patient, trying to enable the patient to notice and to reward himself even for small progress. The program also targets high self-demands as a central theme, encouraging the patient to set himself smaller goals and teaching him self-reflection and positive self-reinforcement. In this context, the "Compass" is also suitable for use as a diagnostic instrument. In Fig. 12.6

Kompass

Patient: Anonymus

Date: 20.07.12

Goal: get welfare claim form **Goal attainment:** 100%

Date: 21.07.12

Goal: fill in claim form **Goal attainment:** 100%

Date: 22.07.12

Goal: get up at 06:30 **Goal attainment:** 100%

Date: 23.07.12

Goal: go to sleep no later than 23.30 hrs **Goal attainment:** 80%

Date: 24.07.12

Goal: stay clean **Goal attainment:** 100%

Date: 27.07.12

Goal: stay clean **Goal attainment:** 100%

Date: 28.07.12

Goal: exhaust yourself through sport **Goal attainment:** 100%

Fig. 12.6 Example of a "compass" over seven units

the "Compass" course for a patient over seven sessions is displayed. Here, the goals partially build on each other, something that can be a very useful side effect; the patient thus creates a structure for task accomplishment. In this case, the therapist gives support only insofar as giving assistance with implementation.

Experience shows that over the course of treatment, the patients usually bring their own ideas for very specific goals to the "Compass" session. The task of the therapist is then focused on substantiating the goals or observing adherence to the "Compass" rules (One goal/Positive rule), as well as to positively support the patient in goal achievement and/or the attempt. According to clinical experience, goals such as "remain abstinent" are selected if the patients particularly suffer from craving, e.g. for cannabis, Internet gaming etc. and therefore want to confirm their abstinence goal.

12.18 Effectiveness of the "Compass" in the Treatment of Cannabis and Party Drug Users

In the LVR Clinic's own study, carried out with 104 cannabis and party drug patients within the scope of the "Bonn model—young addiction" treatment program, there were clear indications of an improvement in the effectiveness of inpatient treatment through the use of the "Compass" (Dau et al. 2011). In comparison to the treatment-as-usual control group, significant improvements of medium effect sizes ($d = 0.56–0.70$) were shown for the "Compass" patients in the following areas: interpersonal problems, depression, trait anxiety and general psychological distress. The "Compass" group also demonstrated better clinical outcomes: more improvements above chance were shown in the "Compass" group on the Reliable Change Index (RCI). The RCI equals the difference between pre- and post-test scores, divided by the standard error of the difference. Cut-off scores are established, assigning patients to one of the following categories; recovered, improved, unchanged or deteriorated (Jacobson and Truax 1991). In terms of general psychological distress, more "Compass" patients improved from clinically increased to normal values (Kendall et al. 2004). Moreover, less relapses with cannabis occurred in the "Compass" group than in the group of patients without "Compass", which is a further indication that the "Compass" intervention facilitates self-management abilities.

Despite limitations due to the quasi-experimental and naturalistic study design, these results are encouraging. Based on patient feedback, the clinical experience was perceived positively and the short intervention was well accepted. Here, the overall impression was that the "Compass" program significantly contributed to improving therapy motivation and the compliance with the inpatient setting more generally, among the young (23 years.) target group of cannabis and amphetamine consumers. Difficulties, like low motivation, reactance, and dropping-out of treatment, that often occur in non-specific therapeutic interventions for this target group, were avoided, and thus the risk of untimely termination of therapy was limited (Aden et al. 2011). This last point was particularly instrumental in adapting the "compass" intervention to outpatient and inpatient treatment of problematic Internet use.

12.18.1 Using the "Compass" Intervention in the Treatment of Problematic Internet Use

12.18.1.1 Acute Inpatient Treatment of Patients with PIU

Previous treatment demands in the outpatient department of the LVR Clinic of Bonn referred to all spectra of problematic use of computers and the Internet. In cases where a comorbid severe and pronounced mental health problem, e.g. severe depression, made acute inpatient treatment unavoidable, the inpatient treatment was undertaken within the scope of the "Bonn model—young addiction" treatment offer. Here, experience so far suggests that the integration of patients with problematic Internet or computer use into the group of young cannabis and party drug patients is possible. Joint treatment within the framework of the psychoeducational group, the motivational group based on the concept of Motivational Interviewing (Miller and Rollnick 2002), and the competence group, in which rejection skills, assertive skills and strategies for relapse prevention are practiced, seem sensible from a clinical perspective, for both patient groups. In this case, the developmental-psychology approach of the "Bonn model—young addiction" (Dau et al. 2008) proved very helpful for the integrative treatment of both patient groups. Although the case numbers are insufficient for systematic statistical evaluation, the clinical impression suggests that both patient groups seem to profit from the exchange of similarities and differences in their respective issues. Within the scope of the inpatient treatment, the patients with PIU also participated in "Compass" and no differences between them and the cannabis and party drug patients became apparent. Patients with PIU accepted the "Compass" and benefitted from the Intervention in terms of therapy outcome.

12.18.1.2 Outpatient Counselling and Treatment of Patients with PIU Undergoing the "Compass" Program

The demand for outpatient interventions far exceeds that for inpatient treatment. In addition to the therapeutic interventions already described, some first steps have been made in adapting the "Compass" intervention to outpatient treatment. Unlike the inpatient practice, discussion of the goals and results is not part of daily therapy in outpatient treatment; rather, this is conducted on a weekly basis. Thus, the "Compass" program takes on the character of a therapeutic 'homework' task (Kazantzis et al. 2005). The principle of breaking one goal down into smaller steps remains central to the intervention. These smaller sub-steps of goals are easier to achieve and it is also easier to train patients to reflect on themselves, to formulate positive and realistic targets and to reaffirm themselves, even for partial successes. Although the literature on effective treatments for procrastination behaviour is scarce (Klingsieck 2013), early studies show that, at least for random student samples, the four elements embodied in the "Compass" (see Box 3), have a favourable effect on procrastination behaviour (Pychyl and Binder 2004).

Box 3: The four therapeutic core elements of the "Compass"
1. Self-observation.
2. Segmentation of tasks into small steps.
3. Formulating of positive approach goals.
4. Training of self-rewarding behaviour.

Clinical experience suggests the "Compass" program is both useful and effective in the treatment of problematic Internet use. Acceptance of the intervention among outpatient clients is good and allows for easy dovetailing of the content of individual therapy sessions through continuous work on self-set and self-relevant goals. Patients quickly see tangible progress, which in turn increases motivation for treatment. Besides this, the "Compass" can be used over and over again to practice the formulation of realistic goals and to discuss goal-oriented problem-solving strategies (D'Zurilla and Nezu 2010). Usually, within "Compass", selected goals focus directly on the problematic Internet use and address related problem areas, such as the development of hobbies, improvement of social skills, etc. It should be noted that "Compass" is not a stand-alone intervention, but a program that supplements an over-all cognitive-behavioural therapeutic approach. Below, a case study will be presented that specifically illustrates the significance of the "Compass".

12.18.2 Case Study

Initial Interview:
Mr. B virtually burst into the room, where he finally got rid of his complaints with vigorous urgency and restlessness. In his case, recording of a structured medical history was not possible. Mr. B reported multiple symptoms including stomach and digestive problems, headaches, tremors, dizziness, aggressive impulses, feelings of tension, insomnia with nightmares, and feelings of inferiority. He reported that everything around him became sort of unreal. In subsequent conversations, drive reduction (lack of motivation) and social withdrawal could also be observed. Concentration and memory problems were also reported by the patient, but these could not be objectively established, neither through conversation with the therapist nor by neuropsychological tests. Mr. B worried about a lack of job opportunities and his upcoming change of residence, difficulties in making friends and stressful family situations. He rarely had suicidal thoughts. Sometimes he felt hopeless and developed thoughts of meaninglessness of life.

Mr. B stated that he considered himself to be addicted to computer games. He stated that he had a strong desire to play these games and described unsuccessful attempts to control his own behaviour. A neglect of leisure activities, friends and own health due to an unbalanced and irregular diet and lack of sleep were furthermore reported.

12.18.2.1 Diagnostic Assessment and the Role of Cognitions

Following a structured clinical interview (SKID-I; Wittchen et al. 1997), the symptoms reported were assigned to a generalized anxiety disorder. Furthermore, at the start of treatment, both depressive tendencies and PC gaming addiction were encoded as control variables. These additional symptoms, as assessed by the structured interview, come under both the classifications of depression and impulse control disorder, in accordance with the guidelines of the International Classification of Diseases 10 (ICD-10; Dilling et al. 2011).

The depressive and anxiety symptoms named in the initial interview increased over time and crystallized as the main stress factors. In addition to this, the unsuccessful attempts at reducing computer usage also became a therapeutic focus. "*Feeling bad*", something that Mr. B stated over and over again, and the experience of helplessness was compensated through use of computer games. Mr. B stated that his experience while using the computer took his mind off things. He stated that the above-mentioned symptoms returned upon termination of the gaming session and in particular when leaving his flat. Dysfunctional cognitions manifested in statements such as: "If I play, I am safe. As soon as I leave the computer, the bad experience returns and I have to return to the computer".

12.18.2.2 Construction of Change Motivation and Development of Therapy Goals

In order to promote change motivation, the advantages and disadvantages of the online gaming were discussed as well as other problems linked with it.

Box 4: Initiation of motivation for change

Therapist: "You have named several symptoms at the beginning, from which you are suffering, and other problems that may be associated with the PC gaming. In your view, what should be changed in your life in addition to this?"

Mr. B: "I cannot get myself to choose a profession, I am also unsure as to how I can manage to move to another place. I cannot stay in the present apartment. I would like to find more friends and establish contacts with others again."

It can be assumed that the client had conversations about his problems before he turned to the outpatient clinic. The younger clients are, the more likely it is that these talks have taken place with parents or legal guardians. Therefore, such conservations may have assumed the form of reproaches, and were probably perceived as damaging to self-esteem (Miller and Rollnick 2002). Notably, the afflicted patients are often placed in a defensive position forcing them to justify their behaviour and speak in the interest of self-protection.

These problems can be avoided if the therapist shows a value-neutral, interested and open outcome attitude. The therapist asks open questions on the problems and goals for change of the client. The question about the benefits and the advantages the patient got from his computer use is also essential:

Box 5: Analyzing the advantages and disadvantages of the problematic behaviours

Therapist: "You have already mentioned various problems related to PC games. What are the benefits the games had for you?"

Mr. B: "I'm really good at" League of Legends ", they are talking about me in the forums and I am being admired! I also play together with others. Without the other players I would be all alone. When I'm not playing, I'm lonely and I see that I have failed professionally and I am getting nothing done. Then I feel dizzy and I get uneasy."

Upon determining the problems and goals that are important for change motivation, the behavioural patterns leading to the problematic behaviour were identified. Here, the patient quickly names the computer gaming itself. In doing so, he enters a voluntary commitment and can give up justifying his game play, as may have been necessary previously in conversations with parents, teachers or friends. It is useful here not to immediately judge the behaviour as "bad", "wrong" or "clumsy", but to look at it as a means of preserving self-worth. In other words, it is important to be open and to explore the function the behaviour serves for the client. Mr. B had independently come to recognise how his gaming served the function of fear regulation, but additional reasons soon became clear; in addition to anxiety regulation, Mr. B's gaming served to fulfil a need for admiration, self-esteem and belonging (see Box 5).

Here, basic motives become clear, such as the need for approval, closeness, and self-preservation. Once these have been identified, the convictions and beliefs held in relation to the problematic computer usage could be approached.

12.18.2.3 Modification of Dysfunctional Cognitions and Activation of Resources

Dysfunctional cognitions are often observed in Internet addiction. These manifest in statements such as, "without success in the game I would be a complete loser. The only possibility to have regular contact with others is through the game.

Being average is not enough, one must belong to the best." In cognitive therapy such dysfunctional beliefs are modified, by arguing against them (Beck et al. 2001). For example, both therapist and patient examine what reasons speak in favour of—and against—these convictions. Through the therapeutic process, Mr. B became aware that some statements are indeed changeable (see Box 6).

Box 6: Example of the modification of dysfunctional beliefs

Mr. B	"I would be a total loser if it were not for my success in gaming."
Therapist	"Okay, you have success in gaming. Can you imagine any other reasons that would suggest that you could also succeed in other things besides gaming?"
Mr. B	"Well, I've already finished school and in earlier days I was good at football, but that was only in earlier times."
Therapist	"Does that mean you cannot play football anymore and you now have no school graduation certificate?"
Mr. B	"Sure I could still do that, but I am not doing anything."
Therapist	"Hmm. How could your sentence "I would be a total loser if it were not for my success in gaming" be reworded into something else?"
Mr. B	"I could play football and I have a college degree, but I am not doing anything with it, and so how should I say this?"
Therapist	"How could you rephrase it so that the sentence becomes more coherent for you?"
Mr. B	"I am currently leaving much undone, but I managed my high-school diploma."
Therapist	"How are you doing with this sentence?"
Mr. B	"Of course, it already feels better than the loser sentence. This way, it already becomes easier to do something. I can notice something here after all."

Mr. B identified a desire to be active as a football co-trainer as an approach goal (Grosse et al. 2004). He had played a lot of football during his youth and he enjoyed working with adolescents. He already helped out in his home club as a co-trainer. It became apparent that he already possessed the skills needed to be a co-trainer, for example, sporting skills gained during sports lessons at school, and organisational skills gleaned from the planning of Internet games. Interestingly, Mr. B was now able to use the skills he had learnt while playing computer games both in therapy and in his 'real' life.

On one hand a realistic view of present behaviours and skills can be developed through Socratic debate as depicted above in Box 6, and on the other hand the value-neutral therapeutic stance towards computer gaming can be vividly demonstrated. Mr. B perceived the appreciation of those skills he had acquired as an appreciation of himself as an individual, as a result of which the therapeutic relationship and change motivation was additionally reinforced.

12.18.2.4 The Role of Comorbidity—Integration of Fear Treatment

The patient's history strongly suggested that the anxiety disorder had developed prior the problematic Internet use, and had facilitated the development of the gaming problem. To counter the patient's anxiety, extensive psycho-education on anxiety disorder was provided. Behaviour therapy interventions targeting anxiety disorders were incorporated into the intervention (Schneider and Margraf 1998). The "Compass" program was used to facilitate positive reinforcement. Mr. B set for himself small goals to cope with his anxiety, and further increased his expected self-efficacy and motivation to change through the sense of achievement he experienced.

12.18.2.5 Formulation of "Compass" Goals in the Therapy Process

Box 7: Example dialogue on a "Compass" evaluation

Therapist	"What were you working on in the past week?"
Mr. B	"It is important to me to improve social contacts again. On Tuesday I formulated the goal *to write e-mails to friends at school.*"
Therapist	"And how well did it go?"
Mr. B	"I wrote to 6 people and two answered at once, on the same day. On Friday, Jens from secondary school has also written back. That made me happy. I evaluated my goal achievement with 100 % in the "Compass". I also set myself the goal after football on Saturday to watch the sports show with the others in the club. This has also worked!"

As shown in Box 7, it was possible to work on changes continuously so that the patient soon set himself daily goals like "view flats (for moving to another apartment)", "meet my school friend Daniel" and "go to the cinema with Jessica" among other things. A shift of priorities resulted, moving away from Internet-related behaviour toward social activities and other important goals. In this example, it can be illustrated that focussing too early on the obviously dysfunctional computer use may not be necessary and may even endanger the client's progress in therapy. On the other hand, a plan of treatment that is aligned to one's own values and goals, taking into account any comorbid disorders, seems to better facilitate treatment.

In conclusion, thanks to this treatment protocol, Mr. B was able to leave the house and go to football training. Through continuous work on small daily tasks, partly related to coping with anxieties, partly related to goal formulation (i.e. through use of "Compass"), Mr. B developed more functional cognitions. He recognized the changeability and controllability of his thoughts, feelings and behaviour.

12.18.2.6 Summary

In the illustrated case, reactance was avoided through a planned and transparent approach (Brehm 1966). Change motivation was created. Early concentration on change can generate fear and can create a danger of not being understood. These aspects seem to be of particular importance when working with younger clients. Here, in addition to building a stable working relationship, the aspect of self-observation needs to be taught. In each case, the therapist must be careful that self-observation is directed toward own behaviour and behaviour change, particularly with younger adolescents. The self-reflection typical for this development phase is often orientated toward one's own perceived deficits, leading to depressive symptoms (Nurmi and Salmeda-Aro 2002). This should not be further reinforced in building up symptom diaries during therapy. Instead, the patient needs to recognize and focus on positive change and how this change can occur. In this example, a sense of control was encouraged, which promoted feelings of self-efficacy. Mr. B was very proud when he understood that his symptoms were closely associated with his anxious thoughts and his fear of failure. This insight helped him to apply for a temporary job. He experienced relief and confidence at his ability to change something. The therapeutic plan was then extended by the patient himself, and the goals of the patient became part of the Compass program (Fig. 12.7).

In outpatient treatment, the "Compass" program leads to better cooperation and improves the effectiveness of the behavioural homework. Systematic investigations of the efficacy of the "Compass" for the treatment of problematic Internet use need to be implemented. Hence, in the LVR clinic, a pilot project for online use of the "Compass" is being planned.

12.19 Effective Therapy Approaches to Problematic Internet and Media Use in General

A review by Petersen et al. (2009) states that an evidence-based recommendation of treatment for Internet addiction is not possible, because studies are largely lacking. Wdiyanto and Griffiths (2006) summarised that most treatment approaches comprised cognitive-behavioural paradigms (regardless of their effectiveness). Peukert et al. (2010) recommend cognitive-behavioural and pharmacological treatment approaches as most likely to be successful. The authors suggest that the inclusion of relatives and family members in the intervention may also be helpful. In a meta-analysis of 16 studies, comprising a total of 670 participants, Winkler et al. (2013) confirmed the effectiveness of psychological/pharmacological treatment. Here, the effect strengths of cognitive-behavioural approaches outperformed other therapy approaches in the reduction of time spent online and depressive symptoms, but the number of studies of pharmacological treatments was very low ($n = 3$). Individual therapeutic approaches proved more effective than group therapy, something the authors explain through increased social anxiety and increased

Kompass

Patient: Herr B.

date: 20.03.13

Goal: max. 2h gaming **Goal attainment:** 100%

Date: 21.03.13

Goal: write an application **Goal attainment:** 20%

Date: 22.03.13

Goal: write an application **Goal attainment:** 100%

Date: 23.03.13

Goal: meet with Daniel **Goal attainment:** 80%

Date: 24.03.13

Goal: day without gaming **Goal attainment:** 100%

Date: 27.03.13

Goal: day without gaming **Goal attainment:** 50%

Date: 28.03.13

Goal: tidy up the flat **Goal attainment:** 75%

Date: 29.03.13

Goal: cooking **Goal attainment:** 100%

Fig. 12.7 Example of an outpatient "compass" run

social isolation, as well as the shortage of social competence, which could be an obstacle for group-therapeutic approaches with this patient group (Winkler et al. 2013, p. 326). Higher effect sizes were found for clients who were female, older, and of North American origin. The latter may be explained by unique characteristics of Asian countries. For example, South Korea has the most developed broadband network of the world, three professional online gaming leagues, and online individuality in the context of social structures is of increasingly high importance. Young people, "who have been brought up within a hierarchical, family focused society find that they are able to act out individualism and socialize independently through the Internet for the first time. Therefore, engaging in psychotherapy for IA might mean losing their individual (online) identity" (Winkler et al. 2013, p. 326).

Further research into this area is necessary. To sum up, it can be stated that Internet addiction seems to be treatable. Cognitive-behavioural approaches could be particularly effective. Further research must elucidate what role pharmacological and a combination of psychotherapeutic therapies play. Also the question of whether group therapy settings or individual therapy settings are more helpful in dealing with Internet addiction needs further investigation, because this could prove very significant. A temporary combination of the two settings might be advisable here, where initially one begins with individual therapy and progresses to a group setting later in the treatment.

References

Aboujaoude E, Koran LM, Gamel N et al (2006) Potential markers for problematic Internet use: a telephone survey of 2.513 adults. CNS Spectr 11:750–755

Aden A, Stolle M, Thomasius R (2011) Cannabisbezogene Störungen bei Jugendlichen und jungen Erwachsenen. Sucht 57:215–230

American Psychiatric Association (2013) Diagnostic and statistical manual of mental disorders: DSM-5, 5th edn. American Psychiatric Association, Arlington

Bakken IJ, Wenzel HG, Götestam KG et al (2009) Internet addiction among norwegian adults: a stratified probability sample study. Scand J Psychol 50:121–127

Bandura A (1977) Self-efficacy: toward a unifying theory of behavioural change. Psychol Rev 2:191–215

Batthyány D, Pritz A (2009) Rausch ohne drogen. Springer, Vienna

Baumeister RF, Bratslavsky E, Muraven M, Tice DM (1998) Ego depletion: is the active self a limited resource? J Pers Social Psychol 74:1252–1265

Beck AT, Wright FD, Newman CF (2001) Cognitive therapy of substance abuse. Guilford, New York

Brehm JW (1966) Theory of psychological reactance. Academic Press, New York

Brendl C, Markman AB, Messner C (2003) The devaluation effect: activating a need devalues unrelated objects. J Consum Res: 29:463–473

Bundesamt für Statistik, Schweiz (2010) Informationsgesellschaft- Indikatoren- Haushalte und Bevölkerung - Internetnutzung. Available via: http://www.bfs.admin.ch/bfs/portal/de/index/themen/16/04/key/ap-proche_globale.indicator.30106.301.html?open=5,1#1

Caplan SE (2010) Theory and measurement of generalized problematic Internet use: a two-step approach. Comput Hum Behav 26:1089–1097

Christakis DA (2010) Internet addiction: a 21st century epidemic? BMC Med 8:61

Cloninger CR, Svrakic DM, Przybeck TR (1993) A psychobiological model of temperament and character. Archives of General Psyiatry 50:975–990
Csikszentmihalyi M (1997) Finding flow: the psychology of engagement with everyday life. Basic Books, New York
Czaika S (2011) Internetnutzung in privaten Haushalten in Deutschland. Ergebnisse der Erhebung 2010 In: Statistisches Bundesamt (ed) Wirtschaft und Statistik. Private Internetnutzung. Zensus. Trendschätzung von Arbeitszeitreihen. Energiestatistik. Kombinierter Verkehr. Seeverkehr. Betriebilche Altersvorsorge. Kaufkraftparitäten. Preise. Destatis, Wiesbaden. Available via: https://www.destatis.de/DE/Publikationen/WirtschaftStatistik/Monatsausgaben/WistaAugust11.pdf?__blob=publicationFile
Dau W, Schmidt A, Banger M (2008) "Bonner Modell" zur integrierten Behandlung der Cannabisabhängigkeit: Umsetzung von ressourcenorientierten Ansätzen in ein vernetztes ambulant-stationäres Behandlungsprogramm. In: Fachverband Sucht e. V. (ed) Qualitäten der Suchtbehandlung. Schriftenreihe des Fachverbandes Sucht e. V. Kongress Heidelberg. Neuland Verlag-Gesellschaft, Geesthacht, pp 286–294
Dau W, Schmidt A, Schmidt AF et al (2011) Fünf Minuten täglich: Kompass - eine stationäre Kurzintervention für junge Cannabis-/Partydrogen- patienten nach dem "Bonner Modell - Junge Sucht". Sucht 57:203–214
Davis RA (2001) A cognitive-behavioural model of pathological Internet use. Comput Hum Behav 17:187–195
Davis RA, Flett GL, Besser A (2002) Validation of a new scale for measuring problematic internet use: implications for pre-employment screening. Cyberpsychol and Behav 5:331–345
DeShazer S (1989) Wege der erfolgreichen Kurztherapie. Klett, Stuttgart
Dilling H, Mombour W, Schmidt MH (2011) Internationale Klassifikation psychischer Störungen. ICD-10 Kapitel V (F). Klinisch-diagnostische Leitlinien 8, überarbeitete Auflage, Huber, Bern
D'Zurilla TJ, Nezu AM (2010) Problem-solving therapy. In: Dobson ED (ed) Handbook of cognitive-behavioral therapy, 3rd edn. Guilford Press, New York, pp 197–225
Eichenberg C, Blokus G (2010) Cybersexsucht: Epidemiologie, Diagnostik, Ätiologie und Therapie - Ein Überblick zum Stand der Forschung. Psychologie in Österreich 2(3):142–154
Eidenbenz F (2011) Wenn Verhalten zur Sucht wird. Sucht Magazin: 4–12
Ellison NB, Steinfield C, Lampe C (2007) The benefits of facebook "friends" social capital and college students' use of online social network sites. J Comput Mediated Commun 12:1143–1168
Ferguson MJ, Bargh JA (2004) Liking Is for Doing: the effects of goal pursuit on automatic evaluation. J Pers Soc Psychol 87:557–572
Fishbach A, Shah JY, Kruglanski AW (2004) Emotional transfer in goal systems. J Exp Soc Psychol 40:723–738
Gentile DA, Choo H, Liau A et al (2011) Pathological video game use among youths: a two-year longitudinal study. Pediatrics 127:e319
Griffiths M (2000) Internet addiction—time to be taken seriously? Addict Res 8:413–418
Grosse Holtforth M, Grawe K (2004) Inkongruenz und Fallkonzeption in der Psychologischen Therapie. Verhaltenstherapie 36:9–21
Hagger MS, Wood C, Stiff C, Chatzisarantis NL (2010) Ego depletion and the strength model of self-control: a meta-analysis. Psychol Bull 136:495–525
Haug S, Gabriel C, Flückiger C, Kordy H (2010) Ressourcenaktivierung bei Patienten. Psychotherapeut 55:128–135
Havighurst RJ (1972) Developmental tasks and education. David McKay, New York
Hinsch C, Sheldon KM (2013) The impact of frequent social Internet consumption: increased procrastination and lower life satisfaction. J Consu Behav 12:496–505
Höcker A, Engberding M, Rist F (2013) Prokrastination Therapeutische Praxis. Hogrefe, Göttingen
Horvath CW (2004) Measuring television addiction. J Broadcast Electron Media 48:378–398
Hurrelmann K, Quenzel G (2012) Lebensphase Jugend: Eine Einführung in die sozialwissenschaftliche Jugendforschung (11, vollst. überarb. Aufl.) Beltz, Weinheim

Integral Markt- und Meinungsforschungsges. m. b. H. (2012) Austrian Internet Monitor. Available via: http://mediaresearch.orf.at/index2.htm?Internet/Internet_aim.htm
Jacobson NS, Truax P (1991) Clinical significance: a statistical approach to defining meaningful change in psychotherapy research. J Consult Clinical Psychol 59:12–19
Jones M, Jones E (1999) Mass media. Macmillan Press, London
Kanfer FH, Gealick L (1986) Self-managements methods. In: Kanfer FH, Goldstein AP (eds) Helping people change, 3rd edn. Pergamon, New York, pp 283–345
Kanfer FH, Reinecker H, Schmelzer D (2000) Selbstmanagement-therapie. Springer, Berlin
Katz E, Scannell P (2009) The end of television? Its impact on the world (so far). SAGE, Los Angeles
Kazantzis N, Deane FP, Ronan KR, L'Abate L (2005) Using homework assignments in cognitive behavior therapy. Routledge, New York
Kendall PC, Holmbeck G, Verduin T (2004) Methodology, design and evaluation in psychotherapy research. In: Lambert MJ, Dupper DR (eds) Bergin and garfield's handbook of psychotherapy and behaviour change, 5th edn. Wiley, New York, pp 16–43
Klingsieck KB (2013) Procrastination. Eur Psychol 18:24–34
Köpertz CE, Lejuez CW, Wiers RW, Kruglanski AW (2013) Motivation and self-regulation in addiction: a call for convergence. J Appl Psychol 8:3–24
Köpetz C, Faber T, Fishbach A, Kruglanski AW (2011) The multifinality constraints effect: how goal multiplicity narrows the means set to a focal end. J Pers Soc Psychol 100:810–826
Kratzer S, Hegerl U (2008) Ist "Internetsucht" eine eigenständige Erkrankung? Eine Untersuchung von Menschen mit exzessiver Internetnutzung. Psychiatrische Praxis 35:80–83
Kraut R, Kiesler S, Boneva B et al (2002) Internet paradox revisited. J Soc Issues 58:49–74
Kruglanski AW, Bélanger JJ, Chen X et al (2012) The energetics of motivated cognition: a force-field analysis. Psychol Rev 119:1–20
Latham GP, Locke EA (2013) Goal setting theory 1990. In: Locke EA, Latham GP (eds) New developments in goal setting and task performance. Routledge, New York, p 1–15
Lavoie JA, Pychyl TA (2001) Cyberslacking and the procrastination superhighway: a web-based survey of online procrastination, attitudes, and emotion. Soc Sci Comput Rev 19:431–444
Lei L, Wu Y (2007) Adolescents' paternal attachment and internet use. CyberPsychol Behav 10:633–639
Leven I, Quenzel G, Hurrelmann K (2011) Familie, Schule, Freizeit: Kontinuitäten im Wandel. In Shell Deutschland Holding (ed) 16. Shell Jugendstudie. Jugend 2010. Eine pragmatische Generation behauptet sich, 2nd ed. Fischer Taschenbuch Verlag, Frankfurt a.M., pp 53–128
Li D, Zhang W, Li X et al (2010) Stressful life events and problematic internet use by adolescent females and males: a mediated moderation model. Comput Hum Behav 26:1199–1207
Lim VK (2002) The IT way of loafing on the job: cyberloafing, neutralizing and organizational justice. J Organ Behav 23:675–694
Lin C, Lin SL, Wu CP (2009) The effects of parental monitoring and leisure boredeom on adolescents' internet addiction. Adolescence 44:993–1004
Livingstone S (2009) Half a century of television in the lives of our children. In: Katz E, Scannell P (eds) The end of television? Its impact on the world (so far). SAGE, Los Angeles, pp 151–163
Locke EA, Latham GP (2002) Building a practically useful theory of goal setting and task motivation. Am Psychol 9:705–717
Locke EA, Latham GP (2013) New developments in goal setting and task performance. Routledge, New York
Mann K, Fauth-Bühler M, Seiferth N, Heinz A (2013) Konzept der Verhaltenssüchte und Grenzen des Suchtbegriffs. Der Nervenarzt 84:548–556
McIlWaraith R, Jacobvitz RS, Kubey R, Alexander A (1991) Television addiction. Am Behav Sci 35:104–121
Medienpädagogischer Forschungsverbund Südwest (2010) JIM 2009. Jugend, Information, Multimedia. Basisstudie zum Medienumgang 12- bis 19-Jähriger in Deutschland. Available: http://www.mpfs.de/fileadmin/JIM-pdf12/JIM2012_Endversion.pdf

Medienpädagogischer Forschungsverbund Südwest (2012) JIM 2011. Jugend, Information, (Multi-)Media. Basisstudie zum Medienumgang 12- bis 19-Jähriger in Deutschland. Available: http://www.mpfs.de/fileadmin/JIM-pdf12/JIM2012_Endversion.pdf

Medienpädagogischer Forschungsverbund Südwest (2013) JIM 2012. Jugend, Information, (Multi-)Media. Basisstudie zum Medienumgang 12- bis 19-Jähriger in Deutschland. Available: http://www.mpfs.de/fileadmin/JIM-pdf12/JIM2012_Endversion.pdf

Miller GA, Galanter E, Pribram KH (1960) Plans and structure of behaviour. Holt, Rinehart & Winston, London

Miller WR, Rollnick S (2002) Motivational Interviewing: Preparing people for change, 2nd edn. Guilford Press, New York

Montag C, Jurkiewicz M, Reuter M (2010) Low self-directedness is a better predictor for problematic Internet use than high neuroticism. Comput Hum Behav 26:1531–1535

Montag C et al (2014) Is it meaningful to distinguish between generalized and specific internet addiction? Evidence from a cross-cultural study from Germany, Sweden, Taiwan and China, Asia-Pacific Psychiatry. doi:10.1111/appy.12122

Müller KW, Glaesmer H, Brähler E et al (2013) Prevalence of internet addiction in the general population: results from a German population-based survey. Behav Inf Technol 1–10

Nurmi JE, Salmeda-Aro K (2002) Goal construction, reconstruction and depressive symptoms in a life-span context: the transition from school to work. J Pers Soc Psychol 70:385–420

Petersen K, Weymann N, Schelb Y et al (2009) Pathologischer Internetgebrauch—Epidemiologie, Diagnostik, komorbide Störungen und Behandlungsansätze. Fortschritte der Neurologie und Psychiatrie 77:263–271

Petersen KU, Thomasius R (2010) Beratungs- und Behandlungsangebote zum pathologischen Internetgebrauch in Deutschland. Endbericht. Available: http://drogenbeauftragte.de/fileadmin/dateien-dba/DrogenundSucht/Computerspiele_Internetsucht/Downloads/Beratungsangebote_pathologischer_Internetgebrauch_Endbericht_100531_Drogenbeauftragte.pdf

Peukert P, Sieslack S, Barth G, Batra A (2010) Internet- and computer game addiction: phenomenology, comorbidity, etiology, diagnostics and therapeutic implications for the addictives and their relatives. Psychiatrische Praxis 37:219–224

Postman N (1988) Wir amüsieren uns zu Tode: Urteilsbildung im Zeitalter der Unterhaltungsindustrie. Fischer Taschenbuch Verlag, Frankfurt

Pychyl TA, Binder K (2004) A project-analytic perspective on academic procrastination and intervention. In: Schouwenburg HC, Lay CH, Pychyl TA, Ferrari JR (eds) Counseling the procrastinator in academic settings. Amercian Psychological Association, Washington, D.C

Rehbein F, Mößle T, Arnaud N, Rumpf HJ (2013) Computerspiel- und Internetsucht. Der Nervenarzt 84:569–575

Rehbein F, Psych G, Kleimann M et al (2010) Prevalence and risk factors of video game dependency in adolescence: results of a German Nationwide Survey. Cyberpsychol Behav Social Networking 13:269–277

Ridder CM, Engel B (2010) Massenkommunikation 2010: Mediennutzung im intermediavergleich. Ergebnisse der 10. Welle der ARD/ZDF-Langzeitstudie zur Mediennutzung und -bewertung. Media Perspektiven 11: 523–536. Available: http://www.media-perspektiven.de/uploads/tx_mppublications/11-2010_Ridder.pdf

Rumpf HJ, Meyer C, Kreuzer A, John U (2011) Prävalenz der Internetabhängigkeit (PINTA). Forschungsbericht an das Bundesministerium für Gesundheit. Available: http://www.drogenbeauftragte.de/fileadmin/dateien-dba/DrogenundSucht/Computerspiele_Internetsucht/Downloads/PINTA-Bericht-Endfassung_280611.pdf

Sariyska R, Reuter M, Bey K et al (2014) Self-esteem, personality and Internet addiction: a cross-cultural comparison study. Pers Individ Differ 61–62:28–33

Schneider S, Margraf J (1998) Agoraphobie und Panikstörung. Fortschritte der Psychotherapie, vol 3. Hogrefe, Göttingen

Shapira NA, Goldsmith TD, Keck PE et al (2000) Psychiatric features of individuals with problematic internet use. J Affective Disord 57:434–452

Shell Deutschland Holding (2011) 16. Shell Jugendstudie. Jugend 2010. Eine pragmatische Generation behauptet sich, 2nd edn. Fischer Taschenbuch Verlag, Frankfurt

Statistisches Bundesamt (2011) Pressemitteilung Nr. 474 vom 19.12.2011. 28 Millionen Haushalte in Deutschland haben einen Breitbandanschluss. Available: https://www.destatis.de/DE/PresseService/Presse/Pressemitteilungen/2011/12/PD11_474_63963.html

Steel P (2007) The nature of procrastination: a meta-analytic and theoretical review of quintessential self-regulatory failure. Psychol Bull 133:65–94

te Wildt BT (2011) Störungen von Selbsterleben und Beziehungsverhalten bei Menschen mit Internetabhängigkeit. Sucht 57:17–26

Teske A, Theis P, Müller KW (2013) Internetsucht - Symptom, Impulskontrollstörung oder Suchterkrankung? Eine Übersicht zum Forschungsstand und zu den Implikationen für die therapeutische Praxis. Psychotherapeuten Journal 12:19–26

Thatcher A, Wretschko G, Fridjhon P (2008) Online flow experiences, problematic internet use and Internet procrastination. Comput Hum Behav 24:2236–2254

Valkenburg PM, Peter J (2007) Preadolescents' and adolescents' online communication and their closeness to friends. Dev Psychol 43:267–277

van Eimeren B, Frees B (2013) Rasanter Anstieg des Internetkonsums - Onliner fast drei Stunden täglich im Netz. Media Perspektiven 358–372

van Rooij AJ, Schoenmakers TM, Vermulst AA et al (2011) Online video game addiction: identification of addicted adolescent gamers. Addiction 106:205–212

Wan CS, Chiou WB (2006) Psychological motives and online games addiction: a test of flow theory and humanistic needs theory for Taiwanese adolescents. CyberPsychol Behav 9:317–324

Wdiyanto L, Griffiths MD (2006) Internet addiction: a critical review. Int J Ment Health Addict 4:31–51

Winkler A, Dörsing B, Rief W et al (2013) Treatment of internet addiction: a meta-analysis. Clinical Psychol Rev 33:317–329

Wittchen HU, Zaudig M, Fydrich T (1997) SKIDStrukturiertes Klinisches Interview für DSM-IV. Achse I und II. Hogrefe, Göttingen

Wölfling K (2009) Ambulante Gruppenpsychotherapie bei Computerspielsucht. In: Hardt J, Cramer-Düncher U, Ochs M (eds) Verloren in virtuellen welten: Computerspielsucht im Spannungsfeld von Psychotherapie und Pädagogik Vandenhoeck & Ruprecht, Göttingen, pp 132–150

Wölfling K, Jo C, Bengesser I et al (2013) Computerspiel- und Internetsucht Ein kognitiv-beahviorales Behandlungsmanual. Kohlhammer, Stuttgart

Wood RE, Mento AJ, Locke EA (1987) Task complexity as a moderator of goal effects: a meta-analysis. J Appl Psychol 72:416–425

Zanna MP (1996) Advances in experimental social psychology. Academic Press, San Diego

Chapter 13
The Korean National Policy for Internet Addiction

Young-Sam Koh

Abstract The Korean Government was the first in the world to develop a national policy to tackle the problem of Internet addiction. For this reason, it has received global attention. To combat the problem of Internet addiction, Korea has established specific laws and systems; a governance system is administrated in government offices, and a "master plan" (revised at three year intervals) has been developed. In addition to these measures, many practical spheres of counseling and treatment have also been established. Examples of Korea's efforts to tackle the problem of Internet addiction include the development of the evaluation scale for Internet addiction, an extensive counseling program, and treatment systems linked to hospital care. This study comprehensively outlines Korea's national policy on Internet addiction.

Y.-S. Koh (✉)
Internet Addiction Prevention and Counseling Center (IAPC) of National Information Society Agency (NIA), Seoul, South Korea
e-mail: yeskoh7@gmail.com; yeskoh@hanmail.net

C. Montag and M. Reuter (eds.), *Internet Addiction*, Studies in Neuroscience, Psychology and Behavioral Economics, DOI 10.1007/978-3-319-07242-5_13

13.1 Overview

Having overcome the 1997 Asian financial crisis[1] through a powerful policy of informatization,[2] Korea continues to invest in the development of the Information and Communication Technology (ICT) industry even during the present age of big data and smart computers.

The introduction of such a radical policy of informatization was particularly successful, as it facilitated Korea's joining the circle of developed countries. The dark side of the radical informatization of Korean society, has been the emergence of Internet addiction as a national issue. Korea was the first country in the world to allocate a national budget to tackle the problems of Internet addiction and has established projects for prevention and resolution. Diverse types of training and counseling are offered for the prevention of Internet addiction, which enables hospitals to provide the necessary help for patients. Korea established new laws and systems to maintain these projects. Other countries have yet to set up similar projects.

As the process of informatization accelerates, digital media, including online video games, is an increasingly central part of our global society. In the future, it will be impossible for anyone to live in this global society without using digital media. This means that most countries will face difficulties similar to those currently affecting Korea, particularly in terms of digital media addiction. In what follows, this chapter will introduce Korea's national policy for the prevention and resolution of Internet addiction.

13.2 Status of Internet Addiction in Korea

The Internet Addiction Prevention and Counseling Center of the National Information Society Agency conducts an annual nation-wide survey on Internet addiction, which gives insight into the percentage of the population suffering from Internet Addiction. Presently, the Korean government is developing an appropriate policy in response to Internet addiction based on both the results of the survey and on known characteristics and tendencies of Internet addicts.

[1] The 1997 Asian financial crisis was a period of financial crisis that affected much of East Asia, including Korea. As a means to stabilize the currencies, the Korean government received a loan from the International Monetary Fund (IMF), under the condition that it tighten its financial and fiscal policies, implement a policy for opening to foreign investment, restructure its financial sector and companies, and improve corporate transparency. Moreover, the Korean government stimulated the information industry for an economic revival. In other words, the ICT promotion policy of Korea was used as a tool to revitalize the economy.

[2] Informatization involves establishing communication infrastructure, introducing information technology in the manufacturing sector to enhance productivity, and industrializing games and computer software as a means to become an information-based society.

According to the 2013 survey on Internet addiction (the Ministry of Science, ICT and Future Planning and National Information Society Agency 2013), 7 % of Internet users aged five to fifty-four are classified as 'at risk' for developing Internet addition.[3] When we examine the recent statistics on Internet addiction in Korea, several implications emerge. For example, after the first study was conducted in 2004, the percentage of Internet addicts decreased each subsequent year, despite the increasing number of Internet users. Also the percentage of addicts shows clear differences per age group. For example, the percentage of addicts was 6.4 % among children aged between five and nine, 11.7 % among youths in their teens, 5.9 % among adults aged from twenty to fifty-five, and so on. Of the different age groups, the group of teens showed the highest percentage of addiction. Another important matter to note is that pathological Internet use is being observed at an increasing rate among the middle-aged adult group between 50 and 54, of which, 3.8 % were reported to be addicted to the Internet (the Ministry of Science, ICT and Future Planning & National Information Society Agency 2013).

The fluctuations in the Internet addiction rate among youths represent the primary focus of the Internet addiction debate in Korea. This is supported by the belief that addiction during adolescence could especially harm the foundation of one's life with profound consequences for adulthood (e.g. Anderson and Bushman 2001; Griffiths and Hunt 1998).

The results of the survey by the above mentioned Ministry of Science, ICT and Future Planning & National Information Society Agency (2013) showed that the Internet addiction rate was higher for adolescent males (12.8 %) than adolescent females (10.5 %) and for male middle school students (15.0 %) than female middle school students (9.8 %). In addition, the Internet addition rate was higher for adolescents from single-parent households (12.0 %) and

[3] The various sub-categories of Internet addiction are differentially explained. For example Kimberly S. Young categorizes Internet addiction as computer game addiction, information overload, network obsessive compulsion, cyber relationship addiction, cybersex addiction, and so on (Young 1998). Davis (2001a, b) used the terminology, 'Pathological Internet Use,' instead of 'Internet addiction' and classified PIU into two types: namely, generalized PIU and specific PIU. Generalized PIU occurs due to the nature of Internet itself such as the anonymity of cyber space, convenience of use, communication speed, immediate response of the Internet, etc. Meanwhile, it is worth acknowledging the influence of gaming and content. Many individuals are severely addicted to computer games, due to game-specific features, such as high levels of stimulation, fantasy etc. Among online games, Massive Multiplayer Online Role Playing Game (MMORPGs) act like an 'emergency exit from reality' for those who cannot adapt and socialize. Recently, Montag et al. (2014) found empirically that there is distinction between specific (such as online video gaming, online shopping, online social networking, and use of online pornography) and generalized Internet addiction, as proposed by Davis (2001a). However there was a large correlation between addiction to online social networks (such as Facebook in Germany and Taiwan or QZone in China) and generalized Internet addiction. In Korea, the categories are drawn from practical counseling experience and comprise online game addiction, online obscene materials addiction, online search addiction, online community addiction, online chatting addiction etc.

from double-income households (12.3 %) than the average addiction rate for adolescents (11.7 %). As the number of people with smartphones has increased in Korea, an annual survey has also been conducted to examine pathological use of smartphones. These results show that smartphone addiction rates have been increasing annually, with the highest percentage (11.8 %) reported in 2013. Taking participants' age into account, the percentage of smartphone addicts among the youth group (10–19 years) was three times higher (25.5 %) than that for adults (20–54 years, 8.9 %). Another important finding was that individuals from higher income households are at a greater risk of developing smartphone addition.

13.3 Korea's Response to Internet Addiction

Korea's response to Internet addiction may appear quite remarkable to other countries. The government has been at the forefront of finding a solution to this problem. This differs from the USA and countries in Europe, where universities and research institutes primarily conduct research and investigate treatment efficacy and provision. Korea also differs from Japan, where private hospitals and non-governmental organizations are most active in this sphere. The following section provides an overview on the Korean response to Internet addiction.

13.3.1 Response to Internet Addiction

The Korean response to Internet addiction is unusual in that the government stands very much at the frontline. Even more unique, it is not just one specific bureau or office providing this response; rather, a variety of federal bureaus help to fight Internet addiction.

For example, because the proportion of young people experiencing Internet Addiction is high, the Ministry of Gender Equality and Family is one of the organizations involved in developing the response. The Ministry of Culture, Sports and Tourism which oversees the video game industry, has also been involved in developing this response, because online games are a significant facet of Internet addiction. The Ministry of Education is involved in educating elementary, middle and high school students, who display relatively higher Internet addiction rates. The Ministry of Health and Welfare is also involved, as Internet addiction is associated with psychological mechanisms. Additionally, the Ministry of National Defense, the Ministry of Justice, and the Korea Communications Commission, administer policies to fight Internet addiction. The reason for the involvement of the Ministry of National Defense is reflected by the number of soldiers experiencing Internet addiction, while the involvement of the Ministry of Justice is reflected in the volume of crimes committed by Internet addicts.

Despite the large number of governmental institutions working to combat Internet addiction, it is the Ministry of Science, ICT and Future Planning, which currently plays the key role in this work. Overseeing the ICT promotion policy,[4] the Ministry of Science, ICT and Future Planning considers the Internet addiction policy a specialized field of ICT promotion policy. In accordance with the Framework Act on National Informatization, as the ministry responsible for policy regarding Internet addiction, the Ministry of Science, ICT and Future Planning consolidates the work of other associated ministries on the one hand, and also carries out its work through the National Information Society Agency (NIA).

There has been constant controversy over the efficiency of projects established by the eight ministries to tackle Internet addiction. Arguments abound over whether it would be more efficient for a single ministry to be responsible for the issue. In any case, it is apparent that projects for Internet addiction receive top priority from the state.

13.3.2 Laws and Systems of Response to Internet Addiction

Laws and systems must be furnished if the national policy is to remain viable. The projects tackling Internet addiction designed by the eight ministries outlined above, are administered according to the laws governing each ministry. The competent law for each ministry is outlined below.

As the organization responsible for overseeing informatization and the response to Internet addiction, the Ministry of Science, ICT and Future Planning oversees the 'Framework Act on National Informatization.' As the basic law for the ICT promotion policy of the state, the Framework Act on National Informatization encompasses the vast field of national informatization, and stipulates the establishment of a master plan to mitigate Internet addiction.

The Ministry of Gender Equality and Family administers the 'Juvenile Protection Act.' The purpose of the Act is to protect the youth generation from harmful or inappropriate environments, and protects them from overexposure to the online gaming environment. Based on the Act, a new system entitled 'The Shut-down System' was introduced to prohibit individuals under the age of 16 from playing games between 12 midnight and 6 am. Specialists presently doubt the effectiveness of the Act,[5] and

---

[4] The ICT promotion policy is a national policy implemented to create advanced wired and wireless communication networks for the rapid exchange of information, as well as to establish an e-government, utilize big data, stimulate the software industry, develop the digital contents industry, mitigate the digital divide, and create a sound online culture. The Internet addiction response policy tends to work against the ICT policy, and thus, the latter policy must include the necessary countermeasures.

[5] Many minors currently use the identification numbers of their parents and other adults to play games after midnight. Also, this system does not apply to game sites with overseas servers and smartphone games.

it has been criticized by politicians who support the gaming industry. However the law is still active, owing to the powerful support of many parents.

The Ministry of Culture, Sports and Tourism administers the 'Game Industry Promotion Act.' This Act provides details for the prevention of excessive preoccupation with games. To expand on this, the principle of 'a selective shutdown' needs to be introduced. This system does not rigorously prohibit gaming during the nighttime hours between twelve midnight to six the next morning. Instead, the guardian can forbid the young individuals from playing games at any time during the 24 h of a given day. Thus, compared to the Shut-down System, it enables the individuals to exercise autonomy more freely.[6]

Moreover, as Internet addiction falls within the area of mental health, the Ministry of Health and Welfare is involved in the response through the administration of the 'Mental Health Act.' A very delicate issue has recently emerged in the context of the prevention and resolution of Internet addiction. The Ministry of Health and Welfare seeks to enact a law that will view Internet games as an addiction factor similar to drugs, alcohol, gambling, and so on, which must be resolved. This same law identified the above-mentioned addictions as '4 major addictions' and a national system was established to deal with them in a comprehensive manner. Another law currently awaiting approval specifies that game corporations should provide a substantial monetary donation to fight against Internet addiction, which will be managed by the Ministry of Health and Welfare.

In Korea, there is currently much conflict between representatives of the game industry, the ministries targeting addiction, nongovernmental organizations, and politicians over this legislation. Those who support it view Internet gaming as very harmful, presenting an obstacle to family harmony, school achievement, and so on. In contrast, people who are against these new laws criticize the view that games are as harmful as drugs such as alcohol, due to the lack of adequate research on the subject.

13.3.3 Key Policy per Ministry

For Korea, multiple ministries are charged with responding to Internet addiction as illustrated above. Thus the response system for Internet addiction is not simple.

As the table shows, the Ministry of Science, ICT and Future Planning is the chief overseeing body for Korea. For example, according to the Framework Act on National Informatization, the Ministry establishes a 3-year master plan, which outlines projects to be undertaken by the various ministries, and manages the implementation of the master plan (Koh 2014). The ministry has also established

[6] Guardians of adolescents below the age of 18 can access the homepage of a game that they want to prohibit and demand from the game company technological measures that prevent access to the game from the guardian's home.

an Internet addiction response center (IAPC: Internet Addiction Prevention and Counseling Center) within an affiliated organization—the National Information Society Agency, which aims to train counseling professionals, develop the Internet addiction diagnostic scale ('K-Scale') for Internet abusers (National Information Society Agency 2013), develop counseling programs for various age groups, administer an annual survey on Internet addiction and develop policies based on the results. A new aim seeks to combat and prevent the rising smartphone addiction. A full list of ministries dealing with Internet addiction and their agenda is presented in Table 13.1.

Table 13.1 Key policies per ministry related to Internet addiction problems

Classification	Details of key projects
Ministry of Science, ICT and Future Planning	• General oversight of responses to Internet addiction according to the 'Framework Act on National Informatization'
	• Establishes an Internet Addiction Prevention & Counseling Center (IAPC) in a National Information Society Agency (NIA) which plans and executes national informatization policies
	• Develops human resources for professional counseling such as fostering counseling professionals, administering staff training, etc.
	• Develops Internet addiction diagnostic scale (K-Scale) and counseling programs
	• Conducts an annual survey of Internet addiction
	• Establishes and manages a 3-year master plan
	• Sets up projects to prevent, counsel, and manage outcomes for Internet and smartphone addictions
Ministry of Culture, Sports and Tourism	• Oversees the interventions to fight side effects of Internet games according to the 'Game Industry Promotion Act'
	• Plans for preventative training and campaign of excessive preoccupation with games
	• Develops diagnostic scales to screen game preoccupation users and counseling programs
	• Supports hospital care for excessive preoccupation with games
Ministry of Gender Equality and Family	• Oversees youth protection according to the 'Juvenile Protection Act'
	• Sets up projects for counseling in over 180 youth counseling centers countrywide
	• Administers residential schools for limited term care
	• Provides financial support for medical treatment for those at high risk of internet addiction
Ministry of Health and Welfare	• Research medical scientific facts and develop a treatment model of internet addiction
	• Operating 200 mental health clinics countrywide
Ministry of Education	• Cooperates with various ministries' school-related projects to respond to Internet addiction
Autonomous regional organizations	• Seoul Metro City: establishes and administers 'the I Will Center"
	• Other sixteen regional governments: the Ministry of Science, ICT and Future Planning and NIA establish and administer each response and counseling center local branches

13.3.4 *Professional Institutions for Preventing and Solving Internet Addiction*

In Korea, there are national agencies which are not governmental departments but are operated with funding provided by the government. Civil servants working for government ministries are transferred to different departments or assigned different jobs every couple of years, meaning that they cannot gain the extensive knowledge required to plan and execute policies for highly specialized fields. On the other hand, national agencies are staffed by people with a master's or doctoral degree, who can contribute their expertise in specific areas, and for this reason, national agencies are considered useful by the government.

Korea's policy for Internet addiction cannot be explained without mentioning the National Information Society Agency (NIA, www.nia.or.kr). While the Ministry of Science, ICT and Future Planning is the overarching government body for Internet addiction, the NIA represents a professional institution for national informatization, which plans and executes policies to support the work of the ministry. With its budgets and projects ratified by the National Assembly, the NIA opened the Internet Addiction Prevention and Counseling Center (IAPC, www.iapc.or.kr) in 2002. This is the first of its kind worldwide, in which a state institution countenances systematic activity related to Internet addiction.

The NIA is presently establishing Internet addiction response centers across regional governments. The NIA has established a response center in thirteen regional governments, of a total of seventeen metropolitan regional governments, and the IAPC headquarters directly allocates human resources and project budgets for administration/management. Approximately fifty Internet addiction researchers, counselors, project planners and directors of IAPC are currently active countrywide. Also IAPC administers over three hundred freelance counselors so that the people can readily receive counseling.

There is also another Internet addiction prevention and counseling institution. It is the Korea Youth Counseling and Welfare Institute, affiliated with the Ministry of Gender Equality and Family. The same organization was originally established to analyze—and provide counseling for—youth problems such as depression, anxiety, study-related stress, family relations, friendships, future plans, etc. Administering approximately two hundred branch organizations countrywide, the same institute also works to prevent—and provide counseling for—Internet addiction, as this problem becomes increasingly serious in the everyday life of youths. Hospitals for counseling and care related to game preoccupation have also been established countrywide, for those suffering from serious game addiction or comorbid symptoms such as depression and requiring medication.

13.4 Internet Addiction Response Program in Korea

The Internet addiction response program in Korea is administered according to the project plans of each ministry, with one year as a unit. The plans are developed according to the comprehensive plan to prevent and solve Internet addiction. At the national level, the program is composed of the prevention campaign, and training, counseling, treatment, post-treatment outcome management, and policy formulation. The next section examines the contents of the comprehensive plan, as well as the projects mentioned earlier.

13.4.1 The Second Master Plan to Prevent and Solve Internet Addiction

The Second Comprehensive Plan constitutes the working plan for 2013–2015 (Koh 2014). As the supervisory body for the plan, the Ministry of Science, ICT and Future Planning along with National Information Society Agency, developed this plan after analyzing and evaluating the results of previous projects for solving addiction over the past three years by the eight ministries. The Master Plan's policy objective is to 'foster capacity to regulate Internet use and promote recovery function,' through the plan's total of fifty-two projects.

The key characteristics of this plan are outlined below. First, the danger of addiction as a result of the proliferation of smart phone technology has been sufficiently considered. Second, in divergence with the first master plan implemented to prevent and provide counseling for Internet addiction among adolescents, we are now providing support for all age groups of Internet and smartphone users ranging from young childhood to adults. Third, the plan should extend services to encompass post-treatment outcome management, beyond the existing program of prevention, care, and so on. Fourth, the plan should be tailored to the specific needs of the client. Fifth, the plan should establish a form of governance among the federal government, regional government, and civilians from schools, corporations, NGOs, etc. Sixth, the plan should develop competent well-trained professionals.

13.4.2 Prevention

Prevention is a key aspect in tackling the problem of Internet addiction. With this in mind, special efforts are put into prevention training. For example, the law provides for mandatory training once a year, for prevention of Internet addiction in children, elementary, junior high, and high school youths, and those from public institutions. The content of the preventative training differs for each age group.

For example, the prevention program for teenagers aims to promote self-control. The prevention training also supports school club activities, which help facilitate appropriate Internet use by students.

In cultivating prevention, campaigning is as important as training. A recently launched campaign bears the slogan: 'Smart Off Day.' This is to promote recognition of the hazards of spending a disproportionate amount of time using smartphones. The campaign is highlighted through radio, TV, newspapers, and also social media. Prevention activity also entails a story composition contest for people who have overcome Internet addiction. A collection of their stories has been distributed widely, to raise awareness of addiction and develop confidence that one can overcome it.

13.4.3 Screening and Counseling

Unique to the Korean treatment of Internet addiction, is the development and implementation of an evaluation scale.[7] The K-Scale (Korean Internet addiction screening scale) has been developed for youths, adults, soldiers, and others. The K-Scale is used both as a tool for self-screening and for assessment by a clinician. The scale measures seven auxiliary factors, including; withdrawal, tolerance, disturbance of adaptive function, disturbance of reality testing, addictive automatic thoughts of the Internet, virtual interpersonal relationships, and deviant behavior such as cyber bulling, adolescent violence etc.[8] The K-Scale, rather than merely diagnosing the level of Internet addiction for particular individuals, is a tool that effectively conducts group screening to determine who is at risk of developing Internet addiction.

To establish the power of this measure, several international conferences have been convened, at which specialists from each of the Asian countries, including

[7] The validity and reliability of K-Scale have been supported through the following psychometric measures: (1) Process of the preliminary test: developed the preliminary questions for the survey (110 questions), conducted the preliminary survey (300 subjects), analyzed the questions (test reliability: 0.97), and extracted the questions for the actual survey (64 questions); (2) Process of standardization: conducted a nationwide sampling survey (2,000 subjects), analyzed the questions (test reliability: 0.96), extracted the final questions (40 questions), computed the cut-off score (70T, 63T), and classified the resulting groups (high risk group, potential risk group, non-risk group); (3) Process of validation: conducted a validation study (2,781 subjects), verified the difference among the resulting groups, and verified the model fit ($x2 = 17{,}211.828$, $df = 719$, $p = 0.000$; $GFI = 0.901$, $AGFI = 0.887$; $NFI = 0.887$, $NNFI = 0.871$; $RMSEA = 0.05$). For more details, refer to the copy of the English summary available at http://www.iapc.or.kr.

[8] Meanwhile, as more people spend an increasing amount of time using their smartphones, a separate screening scale for smartphones users—the smartphone K-Scale—has been developed and is also being employed. The correlation coefficient of this scale and the K-Scale for youth is $r = 0.67$ ($p < 0.001$).

Japan, China, Thailand etc., utilized this scale and compared their results.[9] Telephone counseling is another representative component of the counseling project for Internet addiction. The call number, 1599-0075, offers emergency counseling to addicts anywhere across the country. This type of service is particularly difficult to find in other countries.

Group counseling is also available and is provided by professional counselors on-site at schools. The counseling is put in place upon receipt of an application from each school early in the year. A variety of group counseling programs have already been developed, including art therapy. These programs are administered in accordance with approaches to other types of addictions such as gambling, explicit materials, games, etc. The programs also target specific mediums, e.g. games, general Internet, smartphones, etc. Another unique feature of counseling in Korea is its 'Home Visitation Counseling Program.' This project entails home visits to individuals who have cut themselves off from social relationships in the real world, in order to play Internet games at home. The objective of the project is to facilitate their eventual attendance at the counseling center rather than to completely treat them at home.

The development and management of professional human resources for Internet addiction counseling must also be considered when discussing this counseling policy and program. Highly specialized therapists focusing on Internet addiction are needed to turn the government's policy into a success.

13.4.4 Treatment

Among Internet addicts, there is a high-risk group, for which counseling has limited efficacy. Typically this group includes people with comorbid difficulties such as depression, anxiety, ADHD, etc. For assessment with the K-Scale, it is important to distinguish between low- and high-risk persons and those with comorbid mental health difficulties, and to provide hospital care as appropriate.

The government provides for some of the costs associated with hospital care for those requiring medication. Depending on the circumstances, the government offers support of around 400 USD for early medical examination, or provides financial aid toward the cost of one year's hospital care. The government puts those addicts at highest risk in direct contact with local hospitals, to ensure they receive cooperative treatment.

Although the government first established a counseling center in 2002, in-depth research by psychiatrists has only begun recently. Through the development of animal models, psychiatric researchers are unveiling neurobiological indicators for addiction and are distinguishing the neurological characteristics of addicts, pursuing research to investigate their biosocial linkage.

[9] For example, International Society of Internet Addiction (2012), National Hospital Organization Kurihama Medical and Addiction Center (2013), International Society of Internet Addiction and National Information Society Agency (2013) etc.

13.5 Tasks Requiring Solution

13.5.1 Enhancement of the Diagnostic Function of K-Scale

As discussed above, Korea has developed and employed a diagnostic scale for Internet addiction known as the K-Scale. Schools and professional counseling centers utilize this scale for the diagnosis of Internet addicts. Moreover, annual surveys are conducted at a national level to assess Internet addiction rates and to accurately inform policy.

The K-Scale differs from the Young-Scale (Young 1998; Widyanto and McMurran 2004), which has been used in most earlier clinical work and research. The Young-Scale has been widely used and to great effect, however, limitations do apply. Namely, as it is developed from the diagnostic criteria for pathological gambling, the Young-Scale is limited to describing the current state of the client and fails to consider previous difficulties in the client's life (Young 1996; Koh 2007; Davis 2001b). In consideration of these limitations, Korean researchers have developed an alternative scale, which already has been psychometrically validated (Kim 2002).

Despite this, it is necessary to enhance the K-Scale as it currently exists. Although the K-Scale proves effective in screening people with a tendency toward addiction from among the general population,[10] there is an insignificant basis for determining the cut-off score for diagnosis. It would be more useful if its scope could be broadened beyond screening, to encompass assessment of the extent of addiction. Current research efforts aim to further develop the K-Scale to be used in diagnostics.[11] In the future, it would be desirable to establish a basis for the methods of screening, preventing, treating and examining the progress of addiction among risk groups, by identifying the psychosocial factors and protective factors of Internet (game) addiction.

13.5.2 Heteronomous Regulation or Creation of Culture?

Should there be regulations for the prevention and resolution of Internet addiction in Korea? If so—to what extent? From these questions, a controversy springs up between those who agree that regulation is necessary and those who disagree with this view.

[10] On the K-Scale, individuals at a risk of developing Internet addiction are classified into either a high-risk group or a potential risk group. Individuals in the high-risk group exhibit problems associated with mental health resulting from excessive gaming such as anxiety and hostility, and require professional treatment, as they cannot exert self-control. On the other hand, individuals in the potential risk group exhibit similar problems albeit to a lesser extent, but still require professional counseling.

[11] A study under the title, "A Validation Study of K-Scale as a Diagnostic Tool," has already been conducted. There are plans to publish this study in an international academic journal.

Korean parents today encounter repeated family conflicts over children's usage of Internet and smartphones. Within this context, many parents believe that the government should take effective measures against the problem of Internet addiction, rather than leaving it to the family. As a result, the Ministry of Gender Equality and Family introduced the so-called 'shutdown system', which has been supplemented by the Ministry of Culture, Sports and Tourism's introduction of 'the selective shutdown system'. However, many people also draw attention to the inherent difficulties with the present system. There are Assemblymen who wish to abolish the system. The Assembly audit posed many critical questions about the efficacy of the system. These criticisms were supported by the findings of Uk-jun Seong (2013), which also suggested that the system is ineffective.

The key problems are the lack of convincing alternatives for parents of affected children, and how parents are to combat addiction should the current system be abolished. Those who oppose it argue that the cause of Internet addiction is not just any individual game, but comprises a variety of factors including socio-economic factors, family/environmental factors, individual psychological factors, etc. They believe it is overly simplistic to think that addiction can be prevented solely by the prohibition of games in isolation of these other factors. This logic is correct. Yet Korean parents' general stance is unreservedly in favor of state regulation. With this in mind, it is time for the present government to develop more viable policies.

13.5.3 Federal Government or Regional Government Guided Governance?

Who should be responsible for dealing with the problem of Internet addiction is a matter of significant controversy. The federal government is currently the primary provider of solutions to the problem of Internet addiction in Korea. Yet controversy exists over how long this role can be fulfilled by the federal government. This is due to the fact that Internet addiction is an issue that is related to the private lives of people living in local communities. In other words, the characteristics and causes of Internet addiction differ for each community. For this reason, solutions must also be different for each community. With that in mind, it is insufficient to suggest that the federal government establish and carry out one uniform plan to solve the problem of Internet addiction. Outside of the Seoul region, there are currently 16 metropolitan and regional governments directly operating apart from the federal government. Henceforth, regional governments should play the central role in carrying out measures to deal with Internet addiction and operate with consideration for the social welfare delivery systems presently existing in relevant communities.

At the same time, there is also the question of how long the government alone should be considered responsible for acting to prevent and solve the problem of Internet addiction. Currently, ICT companies, private psychological counseling centers, religious organizations and other private bodies have taken interest in this issue and are working to provide solutions. In this context, it will be more effective for a greater number of private organizations including businesses and religious groups, and also the press and NGOs, among others, to work together to solve this issue. Moving forward, regional governments must play a central role in determining the actual conditions of addiction in their region, and those in each region are working to provide solutions.

References

Anderson CA, Bushman BJ (2001) Effects of violent video games on aggressive behavior, aggressive cognition, aggressive affect, physiological arousal, and prosocial behavior: a meta-analytic review of the scientific literature. Psychol Sci 12:353–359

Davis RA (2001a) A cognitive-behavioral model of pathological Internet use. Comput Hum Behav 17:187–195

Davis RA (2001b) Internet addicts think differently: an inventory of online cognitions. Available via: http://www.internetaddiction.ca/scale.html. Accessed 3 May 2014

Griffiths MD, Hunt N (1998) Dependence on computer games by adolescents. Psychol Rep 82:475–480

International Society of Internet Addiction (2012) Internet aAddiction in Korea: development of new IA proneness scale (KS-II), internet addiction in Asia. Paper presented at the 1st international symposium of psychology and medicine of internet behaviors, National University of Singapore, 29 June–3 July 2012

International Society of Internet Addiction & National Information Society Agency (2013) The masterplan of internet addiction in Korea, internet addiction in Asia. Paper presented at the 2nd international symposium of psychology and medicine of internet behaviors, Seoul National University, 17–19 Oct 2013

Kim CT (2002) A Study on internet addiction counseling and the development of prevention programs, 02-GP-11, National Information Society Agency of Korea. Available via: http://www.iapc.or.kr. Accessed 5 May 2014

Koh Y-S (2007) Development and application of K-Scale as a diagnostic scale for Korean internet addiction. Paper presented at the international conference on internet addiction of adolescent, Ministry of Gender Equality and Family, Seoul, 24–25 May

Koh Y-S (2014) The actual example and national policies regarding internet addiction in Korea. Paper presented at the1st international congress on internet addiction disorders, Milan, Italy. 21–22 May 2014

Ministry of Science, ICT and Future Planning & National Information Society Agency (2013) 2013 National survey of internet addiction in Korea. Available via: http://www.iapc.or.kr. Accessed 5 May 2014

Montag C, Bey K, Peng S et al (2014) Is it meaningful to distinguish between generalized and specific Internet addiction? Evidence from a cross-cultural study from Germany, Sweden, Taiwan and China. Asia-Pacific Psychiatry

National Hospital Organization Kurihama Medical and Addiction Center (2013) Development and application of K-Scale. Paper presented at the 2nd international internet addiction workshop in Yokohama, Yokohama Convention Center 29 May-1 June 2013

National Information Society Agency (2013) A validation study of K-Scale as a diagnostic tool. Available via: http://www.iapc.or.kr/info/lib/selectInfoLibRschList.do. Accessed 5 May 2014

Seong U (2013) A study of the effects of the game regulations policy implemented by the Korean government on the game playing time of adolescents. Paper presented at the 1st Korea media panel conference, Korea Information Society Development Institute, 8 Nov 2013. Available via: http://stat.kisdi.re.kr/library/Library_detail2.aspx?Division=2&seq=256. Accessed 10 May 2014

Widyanto L, McMurran M (2004) The psychometric properties of the internet addiction test. Cyber Psychol Behav 7:443–450

Young K (1996) The Center for internet addiction. http://netaddiction.com/Kimberly-young/. Accessed 14 May 2014

Young K (1998) Caught in the net: how to recognize the signs of internet addiction and a winning strategy for recovery. Wiley, New York

Appendix
Neuroanatomy

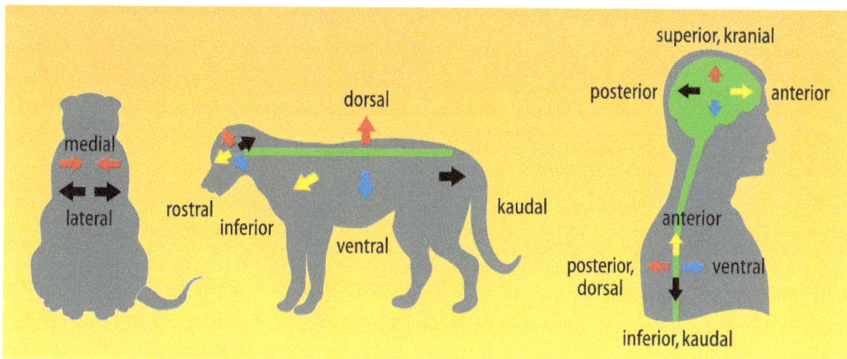

Fig. A.1 Explanation of the orientation in the human brain

C. Montag and M. Reuter (eds.), *Internet Addiction*, Studies in Neuroscience, Psychology and Behavioral Economics, DOI 10.1007/978-3-319-07242-5

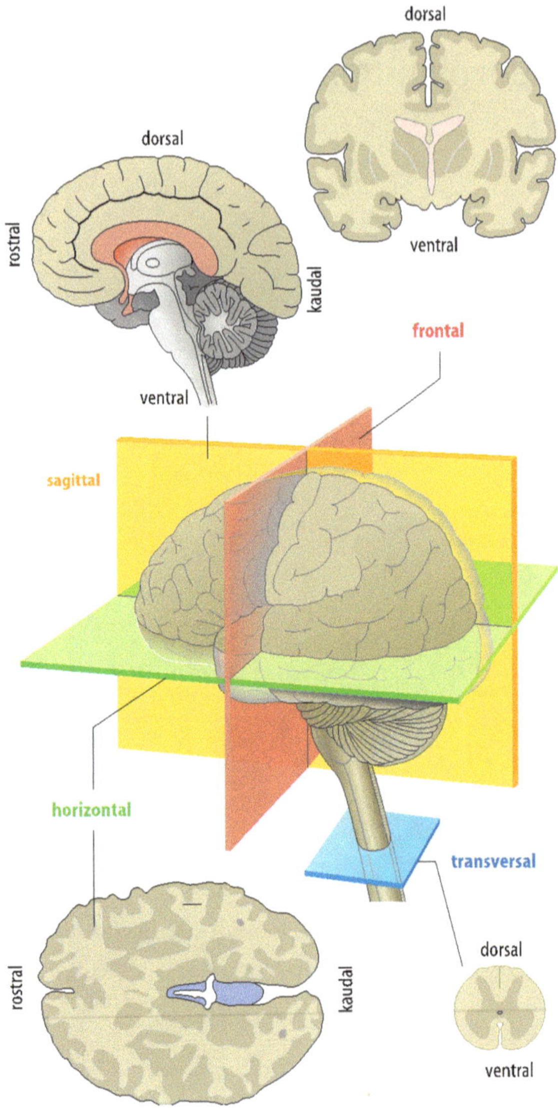

Fig. A.2 Different perspectives on the human brain

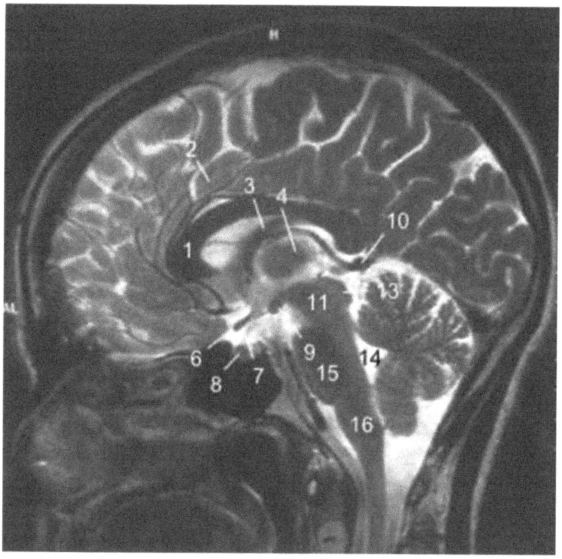

Fig. A.3 Medial sagittal section through the brain (MRI image)

1. Corpus callosum, 2. Gyrus cinguli, 3. Fornix, 4. Thalamus, 5. Chiasma opticum, 6. Infundibulum hypophysis, 7. Hypophysis, 8. Corpus mamillare, 9. Epiphysis, 10. Mesencephalon (midbrain), 11. Aqueductus mesencephali, 12. Lamina tecti, 13. Ventriculus quartus, 14. Pons, 15. Medulla oblongata, 16. Velum medullare superius.

(MRI image from the University clinic in Freiburg, by Dr. J. Klisch, Department of Neuroradiology) (see Fig. A.3).

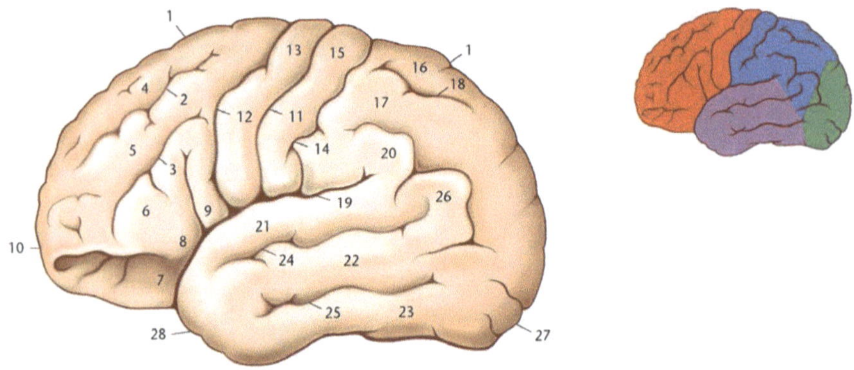

Fig. A.4 Lateral view of the cerebrum

Frontal lobe in red, parietal lobe in blue, occipital lobe in green, temporal lobe in purple.

1. Parasagittal cortical zone, 2. Sulcus frontalis superior, 3. Sulcus frontalis inferior, 4. Gyrus frontalis superior, 5. Gyrus frontalis medius, 6. Gyrus frontalis inferior, 7. Pars orbitalis, 8. Pars triangularis, 9. Pars opercularis, 10. Frontal pole, 11. Sulcus centralis, 12. Sulcus precentralis, 13. Gyrus precentralis, 14. Sulcus postcentralis, 15. Gyrus postcentralis, 16. Sulcus lateralis, 17. Gyrus supramarginalis, 18. Gyrus temporalis superior, 19. Gyrus temporalis medius, 20. Gyrus temporalis inferior, 21. Sulcus temporalis superior, 22. Sulcus temporalis inferior, 23. Gyrus angularis, 24. Lobulus parietalis superior, 25. Lobulus parietalis inferior, 26. Occipital pole, 27. Temporal pole.

(Modified according to Spitzer, in Duus: Neurologisch-topische Diagnostik, Thieme 1990) (see Fig. A.4).

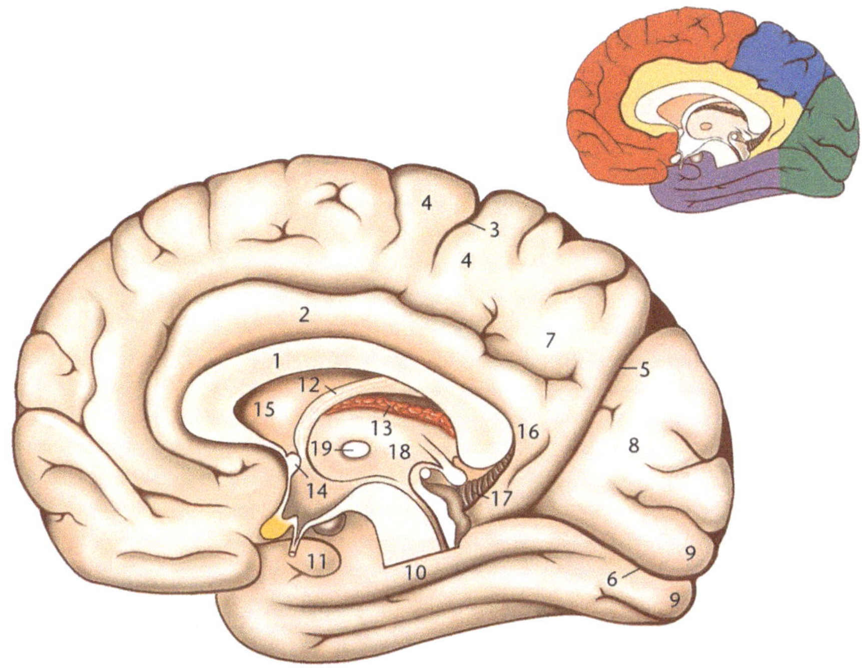

Fig. A.5 Medial view of the cerebrum

Frontal lobe in red, parietal lobe in blue, occipital lobe in green, temporal lobe in purple, Cyrus cinguli in yellow.

1. Corpus callosum, 2. Gyrus cinguli, 3. Sulcus centralis, 4. Lobulus paracentralis, 5. Sulcus parietooccipitalis, 6. Sulcus calcarinus, 7. Precuneus, 8. Cuneus, 9. Visual cortex, 10. Gyrus parahippocampalis, 11. Uncus, 12. Fornix, 13. Tela choroidea, 14. Commissura anterior, 15. Septum pellucidum, 16. Isthmus gyri cinguli, 17. Gyrus dentatus, 18. Thalamus, 19. Adhesio interthalamica.

(Modified according to Spitzer, in Duus: Neurologisch-topische Diagnostik, Thieme 1990) (see Fig. A.5).

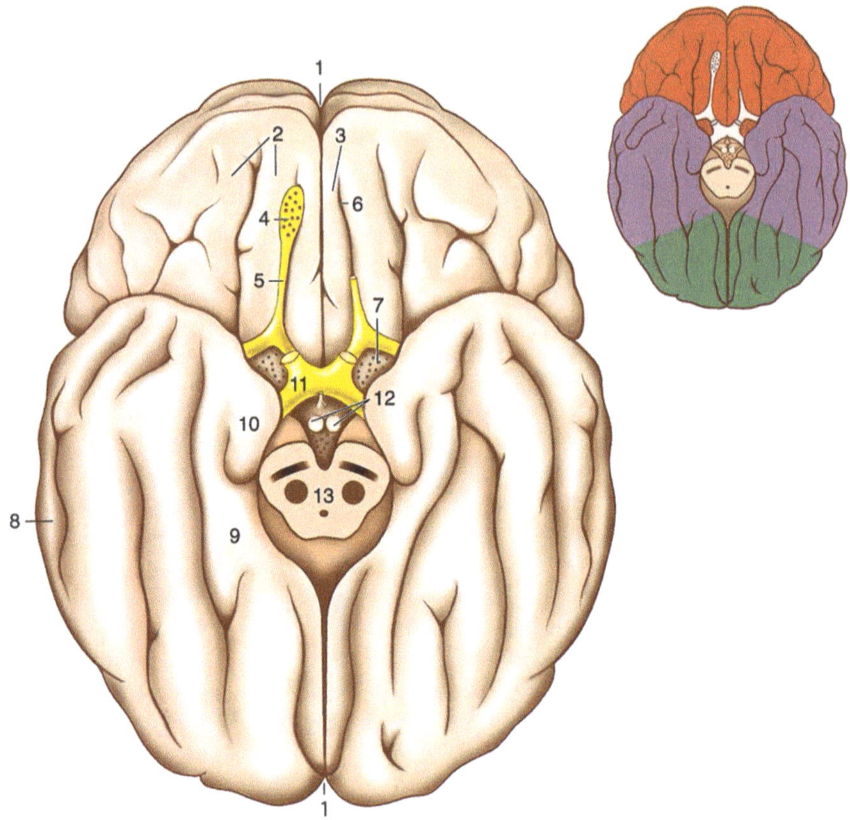

Fig. A.6 Basal view of the cerebrum

Frontal lobe in red, occipital lobe in green, temporal lobe in purple.

1. Fissura longitudinalis cerebri, 2. Gyri orbitales, 3. Gyrus rectus, 4. Bulbus olfactorius, 5. Tractus olfactorius, 6. Sulcus olfactorius, 7. Substantia perforata anterior, 8. Gyrus temporalis inferior, 9. Gyrus parahippocampalis, 10. Uncus, 11. Chiasma opticum, 12. Corpora mamillaria, 13. Midbrain.

(Modified according to Spitzer, in Duus: Neurologisch-topische Diagnostik, Thieme 1990) (see Fig. A.6).

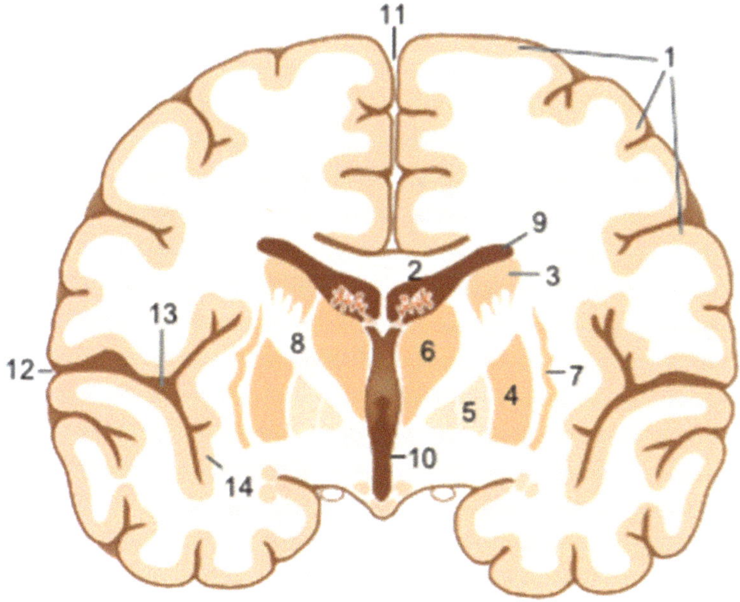

Fig. A.7 The most important inner structures of the cerebrum (frontal section)

1. Cerebral cortex, 2. Corpus callosum, 3. Ncl. caudatus, 4. Putamen, 5. Globus pallidus, 6. Thalamus, 7. Claustrum, 8. Capsula interna, 9. Lateral ventricles, 10. Third ventricle, 11. Fissura longitudinalis cerebri, 12. Sulcus lateralis, 13. Fossa lateralis, 14. Insular cortex (see Fig. A.7).

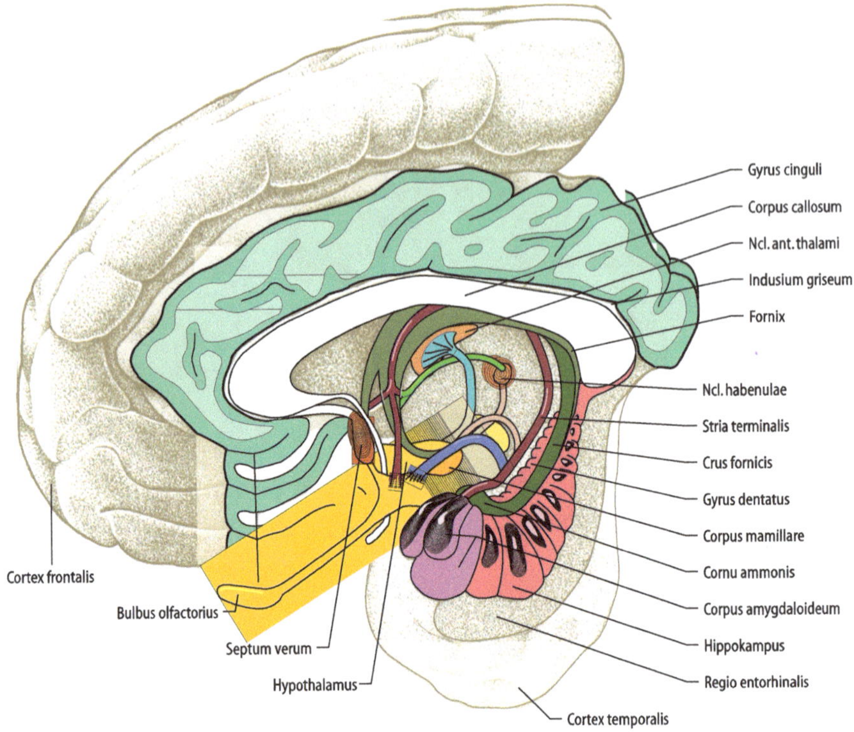

Fig. A.8 View on the limbic system

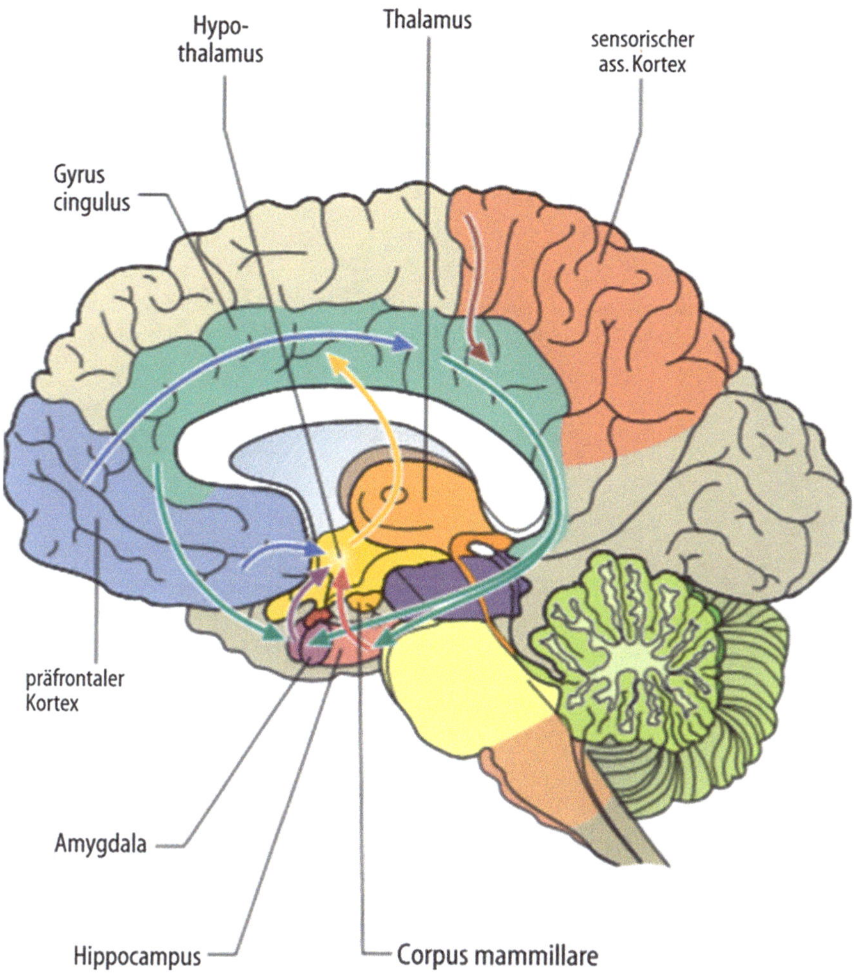

Fig. A.9 Projections of the limbic system

Glossary

DTI Diffusion tensor imaging: By applying this technique it is possible to gain insight into the integrity of white matter tracts by measuring the diffusion of water molecules in the brain. This technique is superior to VBM in investigating white matter.

EEG Electroencephalography: This technique is able to record brain activity from the skull (hence from underlying cortical areas) with a high temporal solution. Of note, compared to fMRI the spatial resolution in subcortical areas is poor.

fMRI Functional magnetic resonance imaging: This technique gives insights into brain activity both in cortical and subcortical areas while doing a specific task. Compared to EEG the temporal solution is much worse. A special version of fMRI is called resting state fMRI, where the activity of the human brain is recorded, while the participant of the study is not engaged in a specific task. This technique gives insights into the default networks of the human brain.

Gene A gene contains the information on how to build a bodily product such as a receptor structure at a neuron in the human brain.

PET Positron emission tomography: While fMRI only gives insights into brain activity, the administration of PET also allows insights into the underlying biochemistry of the human brain.

sMRI Structural magnetic resonance imaging: Through structural brain imaging, researchers gain insights into the structural neuroanatomy of the human brain. Here information on the volume or density of certain brain areas can be provided through VBM. In addition, the investigation of white matter tracts through DTI belongs to the field of sMRI.

SNP Single nucleotide polymorphism: A SNP represents a single exchange of a base at a certain locus of the desoxyribonucleic acid (DNA). SNPs represent a major source for individual differences in the genetic make-up of humans.

C. Montag and M. Reuter (eds.), *Internet Addiction*, Studies in Neuroscience, Psychology and Behavioral Economics, DOI 10.1007/978-3-319-07242-5

VBM Voxel based morphometry: This is a statistical approach often used to obtain information on differences in gray matter volume/density in certain brain areas. Cleary this technique also provides information on white matter. Nevertheless, the VBM approach is not optimal in investigating white matter tracts.

MIX
Papier | Fördert gute Waldnutzung
FSC® C083411

Kolibri 360 GmbH
Ferdinand-Jühlke-Straße 7
99095 Erfurt, Deutschland
produktsicherheit@kolibri360.de